INSECTS, THE CREEPING CONQUERORS

Also by
CARSON I. A. RITCHIE

Making Scientific Toys

INSECTS, THE CREEPING CONQUERORS

AND HUMAN HISTORY

CARSON I. A. RITCHIE

ELSEVIER / NELSON BOOKS

New York

Library of Congress Cataloging in Publication Data

Ritchie, Carson I A
  Insects, the creeping conquerors and human history.

  Includes index.
  SUMMARY: Discusses the effects insects have had on history, art, and
medicine.
    1. Insects—History—Juvenile literature.
  2. Insects, Injurious and beneficial—History—
  Juvenile literature.   3. Animals and civilization—
  Juvenile literature.   [1. Insects—History.
  2. Insects, Injurious and beneficial.
  3. Animals and civilization]   I. Title.
  QL467.2.R57   595.7′06   78–13420
  ISBN 0–8407–6606–8

Published in the United States by Elsevier/Nelson
Books, a division of Elsevier-Dutton Publishing Com-
pany, Inc., New York. Published simultaneously in
Don Mills, Ontario, by Thomas Nelson and Sons
(Canada) Limited.

Printed in the U.S.A. First Edition

10 9 8 7 6 5 4 3 2 1

CONTENTS

INSECTS, THE CREEPING CONQUERORS

ABOUT THIS BOOK

I hope that this book is going to be completely different from any other you have ever read. It sets out to do something that no one else has ever done—to try to explain the definitive effect that insects have had on human history. If you think that this effect has been negligible, please read on. Your views will change.

I have used the word "definitive" deliberately. Two entomologists, scientists who specialize in the study of insects, founded the study of insects in relation to history. One was Frank Cowan, an American, who, with the true detachmant of the scientist, spent the closing year of the Civil War immured in the Library of Congress, writing *Curious Facts in the History of Insects*, a book published by Lippincott in 1865. The other, who combined an interest in entomology with a lifetime of devotion to the study of typhus fever in the laboratory, was Hans Zinsser; in 1935 Zinsser took time off from his studies of typhus to write a medical classic—a biography of the disease and other epidemics called *Rats, Lice and History*. There have been many other writers, before and since, whose interest in entomology has been combined with some other physical science. Thomas Moffet, for example, a physician by profession, wrote the first book about insects printed in English, published in 1634. But, so far as I know, no book about insects in history has ever been written by a historian.

Therefore, I make no apology for drawing different conclusions from those arrived at by past writers on the subject. Entomologists have written about the historical effects of insects as a sort of concluding chapter to the creatures' life cycles. To my mind, the effect of insects on history has been crucial. Take lowly little *Bombyx mori,* for example: the silkworm sparked a long train of events that resulted in the Industrial Revolution and ultimately in the Age of Technology. Yet this is only one aspect of history that can be attributed to insects. There are many others.

I have been forced to do a lot of selecting and omitting to keep this narrative within readable bounds. I have omitted some of the more gruesome consequences of insect-borne disease, for example. I have also said very little about insects in religion. Nevertheless we owe one of the saints of the Church, Saint Malchus, to a colony of black ants.

Malchus was a Mesopotamian monk who toward the end of the Roman Empire was kidnapped by desert Bedouins. Malchus was forced to work as a shepherd, which he did not mind, because he could go on singing psalms and praying while he was tending the sheep and goats belonging to his nomad master. Malchus's master wanted him to marry a girl whom he had kidnapped, but Malchus did not like this idea because he was a monk. Besides, the girl was married already. One day while he was herding his flock Malchus noticed a crowd of ants at work on their ant heap. He thought (as people have thought since) how orderly and businesslike ants are. None of the ants was pushing or shoving, as humans do in a crowd; all were disciplined and patient. It occurred to him that the ants were similar to a busy and orderly company of monks, and he suddenly became very homesick for his monastery. He persuaded the girl to escape with him.

They ran away at night and crossed the Euphrates on inflated goatskins. On the third day of their escape, they discovered that their master and another Arab were following them on camels. They hid themselves among the rocks and undergrowth near the mouth of a cave. Their master, thinking they were hiding in the cave, sent his companion in to bring them out. He went in and did not reappear, so the Arab himself went into the cave. He did not reappear either. Instead there came out of the mouth of the cave a lioness, carrying a cub in her mouth; she bounded off among the rocks. Malchus realized that she had killed their two pursuers. He and the girl mounted the camels and escaped into Roman Mesopotamia.

Another omission in this book is a discussion of the effect that insects have had on human music (though there is a good deal about insects' own music). The most important musical insect is, of course, the tarantula. Strictly speaking, the tarantula is a kind of spider found near Taranto in Italy. It is very large and ugly but probably not venomous. I say "probably" because people react differently to insect poisons, and there may have been a period where a larger than usual percentage of the Italian population was susceptible to the tarantula's bite. In any case, when bitten, the victim sought to shake off the effects by dancing.

"Music," says an eighteenth-century writer called Louis Moreri, "is the unique and sovereign remedy for this disease, for the sick man, violently dancing to the sound of a musical instrument, and even dancing in time, though he may have never learned to dance, expels the disease with the perspiration that breaks out on him." According to tradition the first symptoms of the disease were depression and lethargy, from which the sufferer could be aroused only by music. Whatever the scientific explanation, the tarantula did produce one of the most attractive of all Italian dances, the tarantella.

Nowadays biologists reserve the word "insect" for those members of the class of Arthropoda that have six legs. But since this is a historical study, I have had to define as "insects" all creatures that were once regarded as such. I have included the leeches, which belong to the annelids (jointed worms), as well as spiders and scorpions, which belong to the arachnids (insectlike creatures with eight legs), because in the days before scientific classification, all such animal life was lumped together as insects.

1

PESTS, PLAGUES, AND PREDATORS

Insects are pests only from the human point of view. According to their own lights, insects are merely doing their best. The female mosquito has to have blood before she can lay her eggs. The termites have to have material to build their skyscrapers of clay, and so on. Humans object, because it is their blood that the mosquito drains; or because it is their house that collapses when the termites eat out the foundations. The human beings who have succeeded best in looking at things from an insect point of view are entomologists, the scientists who specialize in insect study.

Entomologists never tire of extolling the virtues of insects. British entomologist Herbert Noyes, for example, pointed out that in 1935 there were 113,034 "beaten, half-starved, verminous, sickly, and deformed children" in England. He contrasted this state of affairs with life inside a termites' nest. The young termites were lovingly cared for by their elders and carefully trained for the work they had to do in later life—even if some of them wanted to play and had to be reproved from time to time.

From most people's point of view, however, insects are far from perfect. Consider *Haemadipsa ceylanica,* a land leech indigenous to Ceylon (or Sri Lanka, to give it its modern name). S. Langdon, a missionary whose garden was a prey to tropical pests, writes:

You will find the grass of my little park too full of life to afford pleasant walking. A few steps on the green carpet will make you acquainted, in all probability, with our notorious land leeches . . . an acquaintance which you will have no wish to renew. They are small to begin with. You may think them scarcely worthy of attention, only about an inch long, and thin as needles; but by the time they have been in your stockings—and they have most insinuating ways—for a quarter of an hour, you will realize that they are capable of wonderful development and enlargement, to which the blood on your leg and stocking will bear witness that you yourself have contributed no small share! In some of our jungles they swarm. During a walk in the jungle, in the district of Avisawella, one day I actually plucked twenty-five of these ferocious little wretches from my legs! Don't venture on the park when the ground is damp. Even when dry it is not a safe place for a picnic. Though the leeches may be absent, the ticks will not spare you. My dog Bob often comes in from his gambols on the park with one of these little, shiny, round, fat creatures hanging on to his ear. The tick is something of a naturalist and takes care to fasten itself on that part of his victim's body which cannot be readily got at.

By the time Mr. Langdon has taken us right around the garden and pointed out the red ants— "take care not to interfere with the red ant *en route,* for if you do, you will doubtless smart for it, I have a vivid remembrance of the savage bites they gave me once" —and shown us the centipedes and millipedes, we are eager to get back inside his house. There the only insects tolerated were the spiders, because they caught flies. Mr. Langdon slept under a mosquito net, although he had been working as a missionary in Ceylon for so many years that the mosquitoes disdained to suck his blood and turned, instead, to young Englishmen freshly arrived in the island, whose blood was richer. Their faces soon looked as though they had measles. Yet the mosquitoes still made life unbearable for him at night, by keeping him awake with their incessant humming.

The dislike of insects—or rather the biting varieties of insects—which Mr. Langdon shows, was shared by all the civilized people of India, even those who lived in rural villages. They too hated the unpleasant leeches and had no desire to venture into the jungle more than they needed to. So the Indian peninsula, because of its abundant insect life, offers a curious anomaly. One of the earliest civilized communities in the world originated in Pakistan, at Mohenjo Dara in the Indus Valley. But although a whole succession of brilliant

civilizations followed that of Mohenjo Dara, none of them extermi-
nated the primitive forest peoples who lived around them.

Both in India and in Sri Lanka, primitive forest dwellers continued
to live almost side by side with civilized men. For these primitives,
the insect-ridden jungle was a perpetual place of refuge from the
agricultural village dweller or townsman, and hence it became a kind
of museum of mankind, in which such people as the Bhils of central
India, culturally akin to the Stone Age, were able to survive down to
our own day. Insects may be said to have preserved them from the
encroachments of their more advanced neighbors.

There is no exact parallel with this situation in North America, but
there are two near parallels. One is the half completion of the
conquest of Yucatán. By 1549 Francisco de Montejo had succeeded
in establishing Spanish rule over barely one half of the peninsula. Not
till 1697 were the last Mayas, the Itzás, subdued. One reason why
Spanish conquest was never carried through completely was the
presence of deadly insect allies of the remnants of the Mayan
population, such as the mosquitoes and chiggers (chigoes) that made
the Yucatán Peninsula so inhospitable.

Another people defended by their biting insects have been the
Eskimos. Centuries ago the Eskimos were driven into the most
inhospitable parts of the Arctic by their hereditary enemies, the
American Indians. The Indians had succeeded in driving them out of
most of the vast territory they must once have possessed in North
America, apparently exterminating them altogether in Labrador. In
1770 Samuel Hearne, who was employed in the service of the
Hudson's Bay Company at Churchill, followed a band of Indians
overland to the mouth of the Coppermine River, where he was a
helpless witness to the massacre of a group of Copper Eskimos by the
Indians. Yet the Eskimos had a powerful ally in *Anopheles
maculipennis*. Great crowds of these mosquitoes collect during the
short Arctic summer, and the accounts of the bites they inflict,
especially on newcomers to their territory, give some indication of
why the frozen wastes of the Canadian barrens and other regions of
the north became the last refuge of the Eskimo. The American Indian
could tolerate most things, but he was sensitive to insect bites. A
Jesuit missionary of the seventeenth century, quoted by historian
Francis Parkman, refers to the fact that Indians always kept a smudge
fire burning in their huts or tents, day and night, to keep off the
mosquitoes.

In addition to preserving people from their enemies, insect pests helped to develop the art of carpet making. As we have seen, nobody in his senses would sit down on the grass of the Indian jungle, no matter how green and inviting it looked. The same caution had to be observed in many other parts of Asia. Grass harbored insects that were not only noxious but deadly, such as the scorpion. Nonetheless, travelers had to rest themselves sometimes on a journey, and they had to sleep outdoors if there was no shelter available. Muslims were required to prostrate themselves on the ground at regular intervals to say the obligatory prayers.

The answer to all these problems was the carpet. This portable "floor" of sheep's wool cut off the occupants from the teeming insect life under their feet. Insects might even be disinclined to crawl across the carpet, because it was made of wool, with all the natural oil left in. Even nowadays, Tartars will tell you that fleas will not hop onto their fleecy wool caftans. Grease is one of the few substances that all insects dislike. It is no accident that the carpet was invented in Assyria. The inhabitants of that country particularly disliked scor-

Travelers have been subject to insect attack since ancient times. Here a French stagecoach is beset by a crowd of cockchafers, a large variety of beetle.

pions, which has always been the most dreaded of all insect pests. The bite of a scorpion need not kill you, but as with all venomous animals, the bite is extremely painful. Carpet weaving spread all over the world, even to countries such as England, where it *is* safe to sit down on the ground. (This is one reason why the Cult of Nature, as exemplified by poets like William Wordsworth, arose in England and not somewhere else.) Saddlebags and saddles were often made of carpeting material. All a traveler needed to do was unharness his mount, and he could sit comfortably, or even sleep, on his saddle and saddlebags.

There was one other important cultural consequence of insect pests. They helped to bring about that fondness for dancing and music which is so characteristic of Africans. This at least is what is suggested by a seventeenth-century French explorer of Africa, Louis Moreau de Chambonneau. Chambonneau visited Senegal between 1674 and 1676. In his account of that trip he gives the following description of mosquitoes:

Among the insects of the country it is surprising how those which we call "mosquitoes" harm and pester men to such a degree. They are a kind of gnat which consists of almost nothing but legs and sting. They are not seen during the day, but as soon as the sun has set you see them coming out of holes in the ground, or from trees and plants, to torment us, during which time, attacking in large numbers, they give us no peace, by their continual humming of: *"Cousins, cousins, ins, ins, ins,"* and biting with their sting so strongly that very often one does not know whether one ought to laugh or cry at the itching which follows subsequently, and which one cannot prevent oneself from scratching in such a way that next morning one has nothing but blisters on one's face, hands and all over one's body where they have bitten, so much so that one is unrecognisable. They come in the rainy season and do not go back again until it is over. I believe that during this time the Senegalese would buy oil of mosquitoes very dearly. Failing this the best remedy they find is to shut oneself up well in a tent bed (if one can sleep in this fashion, because one is stifled under it). As for myself I could not sleep there. After having had a hard struggle inside it against the heat, and not being able to resist, I got up to go for a walk to one of the ends of the island, at the dead time of midnight, when there were plenty of other whites who, like me, defied the linen curtains. The negroes have not got a less sensitive skin than ours to the mosquitoes. Mosquitoes force them often enough to leave their huts, this makes them come together to dance and sing all night.

How extraordinary it would be if Louis Chambonneau's surmise was correct, and that the real origin of African drumming and dancing, which has done so much to enrich the world culturally, lies in this very unpleasant insect! Certainly it is better to get up and do something if you are bothered by mosquitoes at night than just to lie in bed.

Chambonneau, you notice, speaks of getting into a "tent bed" (a four-poster) and drawing the curtains around him. This was not a very efficient kind of protection. A much better deterrent to the mosquito was the mosquito net, which was a coarse fabric suspended from a frame above the bed, so as to protect the occupant from all flying insects. The mosquito net may have been an African, or at least an Egyptian, invention. The Latin poet Horace wrote a poem about the battle of Actium, which was fought in 31 B.C. between Cleopatra and Mark Antony, and Octavian. "Amidst the warriors' standards, oh sight of shame," he writes, "the sun looks down on a mosquito tent!" The tent in question was presumably Cleopatra's. Even nowadays travelers in Central Africa sleep under a tent of muslin. The mosquito net was introduced into Japan by the Portuguese and Spaniards.

In the early sixteenth century, before the introduction of these nets, the only defense the Japanese had against the mosquitoes was lighting fires of green wood—a cure that was almost worse than the disease. Once the clever Japanese had seen a mosquito net, however, they immediately began to make their own. Smudge fires of green wood disappeared, and the mosquito net became so much a part of the Japanese way of life that a disconsolate widow wrote a poem saying that her net was much too big for her, now that her husband was gone.

Curiously enough, the mosquito net never became widely used in Africa, although it must have been introduced there at one time by the Portuguese colonizers. Why did Africans fail to adopt this useful deterrent to being bitten while in bed? It is difficult to say. Orthodox Muslims may have regarded mosquitoes as a plague sent from Allah. One desert Christian, a monk who had killed a mosquito that alighted on his arm, felt that he had given way to anger and impatience. As a penance for his sin, he went and stood in a swamp till he was so bitten by mosquitoes that he was unrecognizable. His fellow monks only knew him by his voice.

A much more efficient deterrent to mosquitoes, and especially flies, was metal window screening. Although invented in Germany in the 1660's, this "wire cloth" never became widely used in Europe, but to American housewives it was a godsend. First marketed in the 1870's, it soon became so popular that women could consider it a greater contributor to human comfort than central heating, electricity, and refrigeration. Ultimately it was to be window screening that would conquer the yellow-fever mosquito and make the construction of the Panama Canal possible.

Until comparatively recent times, men have shown themselves singularly unwilling to do anything about the insect pests that plagued them. But they have sometimes made use of them. Consider the raja of a small Indian state back in the days of the British rule. The raja had been misbehaving in various ways—holding a haremful of women whom he had kidnapped, among other things. The government sent an official named Sir Charles Ogilvie into the state to dispossess the ruler, and he duly marched in with a half battalion of Gurkhas and one armored car. Sir Charles soon found plenty of evidence of the raja's misbehavior, including his imprisoned secretary, whom the ruler had suspected, quite rightly, of corresponding behind his back with the government. After six months of this solitary confinement, the secretary sent a petition to his master, complaining that his health was beginning to suffer from this close confinement. Could he either be allowed regular exercise, or be given Enos (laxative salts)? The raja read the petition, and wrote on it with his own hand: "He shall not have Enos, he *shall* have exercise." Thereafter the secretary's jailers had orders to drop into his cell three live scorpions a day—one in the morning, one at midday, and one at night. The prisoner's whole time thenceforth was devoted to catching and killing these dangerous creatures. When Sir Charles had him released, he had caught and killed all the scorpions dropped into his cell, except the one delivered on the morning Sir Charles arrived. The exercise he got scampering after them may have made him think twice before trying to betray his next employer.

One people who did take steps to protect themselves, especially against scorpions, were the ancient Assyrians. They invented the boot. There is no possibility of a scorpion biting you as long as you are wearing boots, for it cannot jump, and many people wear boots in the jungle today just for this reason. However, the boot had only a modest acceptance in the Near East. The Egyptians, among others,

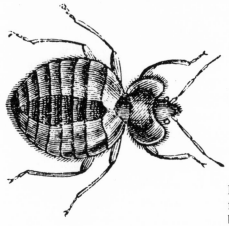

Insect pests were accepted as a fact of life in early days. The bedbug, common in rich man's house as well as poor man's, was regarded as a joke.

continued to go around in open sandals, thus making things easy for scorpions.

Other insects were treated in the same lighthearted fashion. Lice, bedbugs, and fleas were treated as a joke. Rather than change and wash bedding and clothing, men preferred to put up with vermin. John Donne, a famous seventeenth-century English poet, wrote about a flea that had bitten both his mistress and him, so that their two bloods mingled. Theodore Hook was even more explicit. After he had been told by Lady Holland, a society hostess of nineteenth-century England, that he would no longer be a welcome guest at Holland House, he wrote a scornful verse at her expense:

> Her ladyship said, when I went to her house
> She did not regard me three skips of a louse,
> I freely forgave what the dear creature said,
> For ladies will talk of what runs in their head.

Mankind was to pay a heavy price for its easygoing attitude toward vermin. An enormous number of diseases are *insect borne*. Some of the most deadly epidemic diseases in history have been associated with insects—diseases such as the plague, typhus, and cholera. Often, however, the role of the insect in the spread of some diseases must be conjectured rather than proved. Take the Great Plague of 543–544. This plague had begun in Ethiopia. It then passed right through Europe, killing ten thousand people a day in Byzantium alone, but for some reason it stopped at the borders of the kingdom of the Picts, in what is now Scotland. Why should it do that? One reason may have been that the disease was a bubonic plague, spread by fleas. It would appear that the Picts did not harbor so many fleas as their neighbors, the English, the Britons, and the Scots. Perhaps they wore fewer clothes. Perhaps their custom of painting the whole body (their name means "painted men") may have done something to discourage vermin. We do not know. But we do know that the wholesale destruction wrought by the "yellow plague" of 543 A.D. on the very warlike enemies of the Picts gave the Picts a new lease on life. They were able to survive for a few hundred years more than they would probably have done otherwise.

In addition to declining populations, insect-borne diseases cut off important individuals. Alexander the Great, the most successful conqueror the world has ever known, died at an early age with symptoms that are not incompatible with malaria. Alexander had conquered most of the known world. There was very little of it left unsubdued, and mankind was on the brink of having a single world government, liberated from the poison of racialism. Alexander believed that the future for his vast empire lay in intermarriage between the people he had conquered—Persian marrying Greek and Ethiopian Egyptian, till all were one people. But the bite of a mosquito ended this dream.

In the fourteenth century, insect-borne disease delivered a death blow to a way of life. Early in the 1340's most Englishmen lived under the grip of a fuedal society—a system impossible to change. There was little hope of a man's improving his lot or moving up in society. Then a ship put into Weybridge, and some rats skipped ashore. These rats were black rats *(Rattus rattus),* and they carried on them rat fleas *(Xenopsylla cheopsis)* infected with the bubonic plague, a disease similar in nature to the sixth-century one that saved the Picts.

Rats, fleas, and plague had all come a long way. The plague, which became known as the Black Death, had been brought back from the Crimea in 1347 by Genoese soldiers returning home. Plague is endemic in the East but, for causes that are not explained, it is more virulent at some epochs than others. The rat flea and the black rat could catch the plague themselves and transmit it. They did not survive for long with this disease, but they survived long enough to communicate it to other rats and fleas and to humans. During the fourteenth century the pair killed 13 million people in China. Europe lost three quarters of its population—about 75 million people. It was 1348 when the rats came ashore, at Weybridge in Sussex, by means of the ship's hawsers, and started north and east.

The small, enclosed, and self-sufficient communities of medieval England were devastated. Once, researching some medieval papers, I encountered a papal document, dated January 19, 1351, and addressed to the abbot and monks of St. Edmund's Abbey. It authorized the abbot to appoint ten new priests, all of them younger than the canonical age of twenty-five, for all the old priests in the abbey had died, along with perhaps half the population of the house.

This same death toll was reached all over Europe. Few households of the land were untouched. The University of Oxford, to give one example, lost two thirds of its professors and students in 1352. Wrote Giovanni Boccaccio, an eyewitness:

> How many valiant men, how many fair ladies, breakfasted with their kinsfolk and the same night supped with their ancestors in the next world! The condition of the people was pitiable to behold. They sickened by the thousands daily, and died unattended and without help. Many died in the open street, others dying in their houses, made it known that they were dead by the stench of their rotting bodies. Consecrated churchyards did not suffice for the burial of the vast multitude of bodies, which were heaped by the hundreds in vast trenches like goods in a ship's hold, and covered with a little earth.

There were all sorts of reactions in Europe to the Black Death. Some people lost all their religious beliefs in the face of this terrible catastrophe. Others became more religious; people began to talk about reforming the church. In England, a man called John Wycliffe began to put forward what would now be called Protestant ideas.

In 1353, Boccaccio, the Italian writer from whom I have just quoted, published his masterpiece, the *Decameron* ("Ten Days"), created out of plague experiences. A group of people take refuge from the plague and pass the time by telling stories. The tales are very amusing—if a little bawdy—but what interested people about them was that they were all written in Italian, and not, as virtually all books had been in the past, in Latin, the language of learned men and writers. Writing in the vernacular had begun.

Flea-borne plague terrorized Europe for five centuries. The title page of this seventeenth-century account of an Italian visitation depicts much of the horror it invoked.

Shortly afterward, Geoffrey Chaucer decided to try his hand at a collection of stories written in English, and the result was the delightful *Canterbury Tales*, an anthology of stories told by pilgrims riding to the shrine of St. Thomas of Canterbury, in Kent.

Other effects of the plague followed. Seaports learned to "quarantine" ships that had arrived from abroad, especially from countries known to have had outbreaks of plague. *Quaranta* means "forty" in Italian, and quarantine meant that the ship stayed for forty days off the harbor without landing any passengers. If the plague was aboard, then the unfortunate passengers and crew would die from it, but at least it would not come ashore. Medieval men had no idea that the plague was carried by rat fleas, but they tried this idea just to see if it worked, and it did. The quarantine limited plague outbreaks.

In England, the principal changes brought about by the plague were social and economic. Something like half the population died. Whole communities simply disappeared and became the lost villages of medieval England. There was more land than there were farm laborers to work it. Many serfs ran away and achieved the status of freemen, which enabled them to bargain with prospective employers for improved conditions of service. Despite attempts to freeze wages by legislation, they rose by 80 percent during the Black Death century.

Those serfs who were left on the land had to be better treated by their lords, otherwise they would run away, too. Many serfs were set free. By the eighteenth century there was only one serf left in England, a man called Bodo. Bishop Berkeley of Durham used to drop in on Bodo and try to persuade him to become free. Bodo always refused; he knew when he was well off.

With the death of serfdom, the ordinary people in England were free for the first time. They used their freedom to organize two peasants' revolts and to demand that more attention be paid to their problems.

The major epidemic of 1348 was not repeated on a large scale for three hundred years, but the plague did not die out in England altogether. There were outbreaks every year—some large, some small. In London men started publishing a special statistical paper called *The Bill of Mortality,* which listed the number of deaths in certain districts, so that people could keep an eye on the plague and see whether it was on the point of breaking out again.

One reason why the plague did not die out was that the black rat did not disappear. It dug itself in in England, living side by side with humans, a relationship we could well have done without. During the sixteenth century, rats lived underneath the floors of Christ College, Cambridge, where they wrapped themselves up in a warm nest of books and leaves of books that would be worth a fortune nowadays. This rat's nest, discovered in 1911, included an early Virgil, part of a Horace, two leaves of a primer by Wynkyn de Worde, one of the earliest English printers, and a leaf of a book by Caxton, the earliest printer in England. Of course, with this many rats on the premises, it is easy to see why seventeenth-century Cambridge suffered more than Oxford from the plague.

Everybody in England had some rats, however, from the lowly peasant up to the king. There was a royal rat catcher to the king, who

Although the connection between the rat flea and disease was not yet understood, men did know enough to do battle with rats. Here an Elizabethan rat catcher parades with his official banner.

wore a special uniform of scarlet, embroidered in yellow worsted with figures of field mice destroying wheat sheaves.

Nobody did anything about the fleas except to scatter a plant called fleabane in rooms and very occasionally wear fleatraps—a small silver tube filled with mercury—around the neck to kill them. Fleas were always a good joke at home and abroad. So, too, were lice. In France a poet called Pasquier wrote a book of poems called *The Carnival of Poitiers,* about a flea which he had found one morning in the bosom of Catherine de Roches. When visitors toured the French king's galleys in Marseilles, the galley slaves collected all the lice they could find and blew them through paper cones onto the visitors' clothes.

As the tide of plague victims ebbed and rose from year to year, the presence of this epidemic disease continued to modify English society drastically. The more extreme kind of English Protestants, the Puritans, pointed out that there must be some connection between

As the people began to realize that there must be some connection between plague outbreaks and large crowds, many began to flee the towns, taking the infection with them.

plague outbreaks and large crowds. They did not put their point precisely in those words; instead they argued that large assemblies for bear baiting or to watch plays were ungodly and ought to be forbidden by the government, that there should be prayer and fasting instead. Right through the reign of Elizabeth I, Puritans struggled with the Queen and her ministers to get the theaters closed. They were only partly successful. The theaters would close for a time, then open again. For this, lovers of the English language can only be grateful, since it meant that William Shakespeare and his fellow poets of this great literary high point were heard and performed.

In 1665 England was struck by the Great Plague—the most terrible outbreak of this disease since the Black Death in the fourteenth century. Out of a population of 460,000, some 68,596 people were *known* to have died; the actual loss was probably much greater. So many Londoners died that special arrangements had to be made for burying them. They were collected by corpse bearers and hurriedly

A Corpes Bearer

So many people died of the 1665 Great Plague that special officials had to be appointed to carry off the dead.

interred in great pits, still fully clothed. Many complete sets of seventeenth-century apparel have been recovered by archaeologists from these pits. Londoners fled the city, taking the infection with them, so that it devastated the whole country. Eventually, in September 1666, a blessing in disguise saved the city. A great fire broke out and burned for four days, during which 89 churches and 13,200 houses were destroyed—along with a large population of rats and rat fleas. The Great Plague was at an end.

One effect of the Great Plague of 1665 was to change Englishmen's ideas about emigration to the colonies in North America.

Hitherto, people in England had believed that the country was overpopulated. There were too many Englishmen, they thought, and the best way of getting rid of the surplus numbers was to ship undesirables to America. Some of the first American settlers had not wanted to leave England at all—they had been marched aboard ship in handcuffs for the voyage, shipped over by the government to promote the formation of colonies and the foundation of new settlements. Then, after the Great Plague, English interest in North American colonization dropped somewhat. How could England be so overpopulated, people asked themselves, when 70,000 people had perished in the capital alone? The newly founded American colonies were left to grow up by themselves and developed into independent, individual units, where local ideas about freedom prevailed. When the eighteenth century came, America was ripe for independence.

The stoicism with which the English usually bear their troubles was manifested in 1665. There were no riots and not even a great deal of looting. On the European continent, things were different. In 1656 an appalling outbreak of the plague occurred in Naples, where 300,000 people died in five months. Dreadful scenes took place. The survivors accused the Jews of having caused the outbreak by poisoning the wells and attacked them brutally. Another symptom of mass hysteria was the dancing mania. Someone would begin to dance of his or her own accord; immediately everyone else would join in—even old men and women who had not danced for years—and they would keep on dancing until they dropped with exhaustion.

Not only do insect-borne diseases disrupt societies, they destroy the effectiveness of that specialized kind of society known as an army. Armies have been halted in their tracks again and again by the outbreak of epidemics that they were powerless to combat, and which inflicted more casualties on them than any enemy could. Time and time again conquerors have been stopped cold by the onslaught of disease. When the Assyrian Sennacherib was besieging Jerusalem in 701 B.C., his hitherto invincible army was turned back by a plague, and this experience was to be repeated in war after war throughout history, now by one disease, now by another.

Usually the soldiers of an invading army, being strangers to the region, have no resistance to the local disease, whereas the inhabitants, through long exposure, have developed some immunity. Consider the Crusades. Here great armies of enthusiastic European knights and foot soldiers simply melted away in the face of local diseases carried by insects such as the blowfly and housefly—usually dysentery. Dysentery caused such havoc among the Crusaders that the vast hosts which set out for Palestine were reduced, in the course of a single campaign, to mere handfuls.

The grip of insect-borne diseases on an army has seldom been better portrayed than in the first modern biography, the *Life of St. Louis,* written by Jean de Joinville. Louis, King of France between 1226 and 1270, twice invaded Egypt in order to relieve the pressure on Outremer—as the Crusaders' territory was called—in what is now Israel. Joinville portrays the whole French army in a state of defeat and collapse, occasioned not by the attacks of the Saracens of Egypt but by the onslaught of dysentery. With the clinical precision which makes his book so interesting, Joinville portrays the condition of the king, who suffered recurring attacks of dysentery. So frequent were his visits to the closestool that he had his breeches cut open.

It was not necessary to leave Europe to see whole armies prostrate with disease. One of the unhealthiest areas on the continent was Rome. The drainage system of the Roman Empire had been allowed to go to ruin. Extensive marshes, which had been present even in ancient Roman times, had spread. At some point in history these marshes had become the breeding ground for the *Anopheles* mosquito, which transmits malaria to man. Kipling's observation "The female of the species is more deadly than the male" is certainly true of anopheles, for only the female bites. She prefers to have a drink of blood before she lays her eggs. (In a pinch, she will drink fruit juice, but like many human beings, she never drinks the soft stuff when there is any of the hard to be had.) She injects her human victim with her proboscis, which is like a hypodermic syringe, and draws out blood. If she is a malarial vector—that is, one carrying the parasitic protozoan which causes the disease—she infects her victim. The results are regularly recurring fever, delirium, and frequently death.

There may even be some reason to believe that the well-known features of Uncle Sam show some of the characteristics of the malarial sufferer. Uncle Sam wears a goatee beard, suggesting he lives in a rural area, and his clothes date him to the 1830's. At that time the Northern and Southern states alike were ravaged by malaria (there was a particularly bad area of infestation in New Jersey). Possibly the prototype for Uncle Sam had acquired his hollow-cheeked leanness from regular attacks of this terrible disease.

Rome was only one of the regions that suffered from regular infestations of malaria. But the Eternal City was the center of the medieval world and held a special place in men's minds. Again and again the Holy Roman Emperors of Germany and Austria strove to

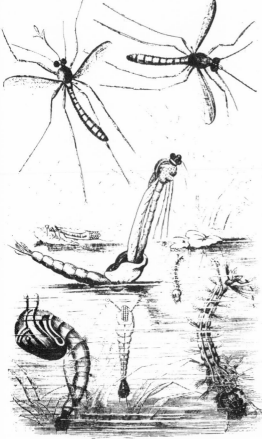

The *Anopheles* mosquito, carrier of the dread malaria, breeds in swamps and marches. In medieval Rome, where the drainage system of classical times had been allowed to decay, malaria thrived.

control Rome and with it the papacy. Armies of stalwart knights would descend from the Alps and besiege the city. There they would be attacked by the swarms of mosquitoes that rose from the Pontine Marshes south of Rome, become infected with malaria, and succumb to the disease, while the besieged army of papal supporters, being made up of local knights and soldiers, who had some immunity to the disease, remained relatively unaffected. The besiegers would be forced to abandon the siege, bury their dead, and march away, and again Rome would have justified its title "Eater of Men."

Not just districts, but whole continents were protected from outside invaders by the mosquito. Take Africa, for example. Some places in Africa are much more malaria free than others. South Africa is mostly immune. Other parts, like Malawi, lie very high up, where mosquitoes appear in smaller numbers than along the great lakes and rivers. But Central Africa is a malarial hotbed.

Africa has always been a tempting prize for outside invaders, but really serious attempts to control the continent began only after the Congress of Berlin in 1880, when the European nations began the "scramble for Africa." Earlier European settlements in Africa before had been, for the most part, dotted along the coast, and the Europeans had settled there in very small numbers. More successful were the Arab invaders of East Africa, such as those who founded the Zenj Empire around Zanzibar. They realized that Africans had some immunity to malaria and, although some of them did contract the disease, they frequently recovered from it. These Arab settlers could not immediately acquire immunity themselves, but they could marry local girls so that their children would be born with partial immunity. The Europeans, on the other hand, kept themselves very much to themselves—and found that they could not bring up white children in the tropics. Thus Africa remained largely unsettled by whites.

The Africans were to pay a terrible price for their semi-immunity to malaria. Unable to make much of Africa itself, the European settlements along the coast exported immunity from malaria by sending cargoes of slaves to the West Indies and to America. The resistance of the Africans to malaria enabled them to survive the sinister "seasoning sickness" that carried off so many white settlers in their first summer.

Early American planters had no particular desire to employ black slaves in the plantations, preferring white indentured labor. The slave was a slave for life, whereas the indentured laborer was released at the end of a certain period of time, usually seven years. But this was a distinction without much difference, since many planters starved or beat their white servants to death before their time was out. Sometimes white slaves did not bear the expense of shipping out, for they often died soon after they arrived. Blacks were a better bargain. So although many American colonists—like General Oglethorpe, the founder of Georgia—would have preferred to do without Negro slaves altogether, they became convinced that they could not. If they wanted to settle in the South, they had to have black slaves to work their plantations. So slavery became an American institution— thanks to the *Anopheles* mosquito.

Many missionaries in Africa, who wanted nothing more than to convert the Africans and educate them so they could cultivate their fields more efficiently or take up peaceful commerce (instead of the destructive slave trade) either died of malaria or were forced by attacks of sickness to return home.

Knight Bruce, the first Bishop of Mashonaland in Rhodesia, had to return home because of malaria. Bishop McKenzie, leader of the Universities' Mission to Central Africa, died of malaria when his canoe was upset in the river and the quinine (an antimalarial drug) was lost. Later the whole mission moved back to Zanzibar because of sickness.

Dr. David Livingstone, the most famous of all African explorers, died of malaria with his work still unfinished. Years earlier he had seen his wife die of the same disease. Many more British missionaries would have come to work in Africa, but for the disease-ridden bush.

Africa was protected by another fearsome insect guardian, the tsetse fly. This insect spreads sleeping sickness, a most unpleasant disease of the central nervous system, characterized by disturbances in the patient's normal sleep patterns, sometimes profound coma. It attacks cattle and horses as well as men, and has always been so dreaded by invaders of Africa that most tsetse-fly belts have been shunned by everyone. There are, for example, almost no rock paintings in the tsetse areas in Africa, suggesting that the races which made these wonderful paintings avoided the sleeping-sickness regions. Later African invaders usually turned back when they found they had entered a tsetse area. This happened to the great Zulu conqueror, Mzilikazi, who had invaded Makolololand, near Livingstone, in Rhodesia. To the Zulus, sleeping sickness meant death, whereas the loss of cattle meant poverty.

It was unfortunate for the Africans that insect sentinels such as the mosquito and tsetse fly kept out not only invaders but people who wanted to help the Africans as well.

The American continent south of the United States was protected from invaders not merely by the malarial mosquito, *Anopheles gambiae,* but by another insect villain, *Aedes aegypti* (originally called *Stegomyia fasciata)*, which is the vector for yellow fever, or "yellow jack." Yellow Jack was the most dreaded of all American diseases. Ships that had a case aboard were supposed to fly a "yellow jack," or yellow flag. So prevalent was yellow fever in Latin America that an order to garrison a West Indian island was equivalent to a sentence of death. Yellow Jack protected the Spanish American colonies from attack by Britain and North America much more effectively than armies or fleets.

The only serious involvement of Napoleon on the American continent was his invasion of Haiti. The Haitian slaves had revolted against their French masters during the French revolutionary wars, setting up the first black state outside of Africa. Napoleon's brother-in-law, General Charles Leclerc, and a crack French army descended on the island, intending to recover this valuable source of cane sugar for France. The Haitians, led by the black liberators Pierre Toussaint L'Ouverture and Jean Dessalines, combated them bravely, but the real victor in the struggle was yellow jack. Out of an invasion force of 25,000, as many as 22,000 died, including General Leclerc. The French invasion force had to pull out. The Haitians were left to their independence and were thus able to prove to the world that a black state could go it alone.

This defeat had important consequences for America. Napoleon, who had earlier experienced the effects of epidemics in his Egyptian campaigns, was convinced by his Haitian disaster that it would be impossible to maintain a large French army in a warm climate. So, when he acquired Louisiana by overrunning Spain, he decided to sell it to the new United States for a bargain price. With the Louisiana Purchase of April 1803, the biggest land deal in history, America acquired a million square miles of territory at a price of about four cents an acre. The decision to sell was so sudden that the American negotiators were taken by surprise. The purchase, then, was not a triumph for American diplomacy so much as a demonstration of how *Aedes aegypti* had frightened the conqueror of Europe.

However, the United States paid considerably more than four cents an acre for Louisiana. By adding new slave states to the Deep South, it ensured that Southern planters would be more than ever tied to a slave-run economy. The Civil War followed as an almost inevitable consequence of the Louisiana Purchase, and while the two halves of America fought and there was no cotton to be had from the

United States, Russia began to emerge as an industrial state and the provider of cotton for Europe.

Yellow fever continued to be one of the greatest plagues that mankind ever had to suffer. Then in 1881 Carlos Juan Finlay (1833–1915), a Cuban doctor, presumably of Scots descent, began to suspect that yellow fever was transmitted from fever sufferers to healthy humans by a mosquito, which he later identified as *Aedes aegypti*.

It took twenty years for someone to take Finlay's theory of the transmission of yellow fever by the mosquito seriously. Then, in 1900, Walter Reed (1851–1902), a doctor who had been investigating the causes of yellow fever since the late 1890's, arrived in Cuba in charge of the U.S. Army Yellow Fever Board. Reed listened to Finlay's exposition, skeptically at first, then with a dawning conviction that made him decide to try his theory out. He undertook a series of experiments, among them one in which four heroic American volunteers slept in huts, using unwashed bedding, pajamas, and eating utensils of yellow-fever sufferers, but screened from mosquitoes. The results were conclusive: They did not get yellow fever. In 1901 United States military engineers began a campaign in Cuba against *Aedes aegypti* that freed Havana from the disease. The way was now open to build the Panama Canal, a project which had already been attempted by the French under Ferdinand de Lesseps

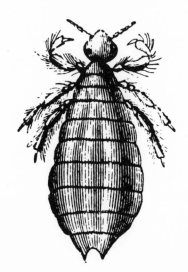

The body louse is the carrier of typhus—often called "jail fever" or "camp fever," because it spread like wildfire wherever large groups of persons were gathered together.

during the 1880's, but which had failed miserably because of yellow fever.

But for the work of Carlos Finlay and Walter Reed—and the courage and determination of their volunteers—the Panama Canal could not have been completed in 1914.

Napoleon's good sense, which had suggested to him the sale of Louisiana to the United States, deserted him when he attacked Russia in 1812. Once more French armies met defeat at the hands of an insect—in this time the louse. Many of the inconveniences of lice is that they can become vectors for a disease called typhus, which is endemic in eastern Europe—that is, always present.

Outside of eastern Europe, typhus had always been associated with overcrowding. It thrived in slums and ghettoes. It flourished among the large crowds present at trials, which then—as now—had a great attraction. In England, typhus was so much associated with jails and trials that it was called the "jail sickness" or "jail fever." The sentence of death pronounced by an English judge with his black cap was no more fatal than the sudden epidemics that struck during assizes (periodic sessions of the superior courts), and which carried off judge, jury, prisoners, and witnesses alike, long before the formality of execution could be carried out on the condemned. The only cure for the "foul air" which was supposed to cause jail fever was thought to be the carrying of a bouquet of sweet herbs. English judges still carry posies of flowers when they arrive for the assizes.

In 1812 Napoleon perpetrated the greatest piece of overcrowding in history. A vast army of half a million men was assembled on the borders of Russia. They marched in, sleeping in crowded camps, carefully huddled together and patrolled by sentries to discourage would-be deserters. They slept in crowded villages—if they were fortunate enough to find a village still standing—and from their Russian captives they caught typhus, which spread like wildfire through the vast army. From the moment Napoleon crossed the Neman River on June 24, 1812, his casualties from the disease began to grow. By late July there were 80,000 Frenchmen on the sick list, almost one sixth of the army. Struck by disease and also by hunger, the French went down like flies before the Russians. Out of an army of half a million men, only 3000 lived to evacuate Russia—and to spread the disease in Napoleon's empire and hasten his downfall.

Paradoxically, even disasters like epidemics can result in good.

Sometimes they can halt wars and revolutions. The revolution of 1848, one of the bloodiest outbreaks of violence in European history, was halted partly by the onset of a new epidemic of typhus. But this does not always happen. Both England and France went on fighting the Hundred Years' War in the fourteenth century, although both had been devastated by the Black Death.

The terrible destruction of life in armies by disease also did more than anything else to improve medical science. It cost so much to train a soldier, feed him, and clothe him that governments slowly began to realize the importance of research. In the eighteenth century, the British government became very worried by the losses through disease in both the Royal Navy and the army. Intense medical research by two physicians, Sir John Pringle and James Lind, did not discover the causes of disease on ships and in barracks, but it did suggest that there could be *some* connection between cleanliness of person, kit, bedding, and living quarters, and immunity from disease. The War Office and the Admiralty, informed of the findings of Pringle and Lind, shrugged their shoulders in disbelief. It seemed very unlikely that there could be any connection between disease and dirt. After all, the habit of not washing was a very old one, ingrained in the European way of life since the fall of the Roman Empire. Nonetheless, anything was worth trying. The order went forth: Henceforth British servicemen were to wash and shave regularly; they were to launder their clothes and bedding and clean up their surroundings—which included discouraging insect life. The English fighting services became a model of cleanliness and health.

The contrast between the English army and other armies at the start of the nineteenth century, when these changes began to take place, was very striking—nowhere more so than when Englishmen met Americans at New Orleans, on January 8, 1815. The English promptly nicknamed the Americans, most of whom wore homespun shirts that were rarely washed, the Dirty Shirts. By contrast, the habit of cleanliness was deeply ingrained into British troops. After the battle, while British and Americans were burying the dead under flag of truce, one British soldier leaped into the grave pit and helped himself to three shirts from his fallen comrades. He thus had one to wear, one to wash, and a third that was drying, and he could discard his own worn-out shirt.

Along with cleanly habits, which spread rather slowly to the whole of industrial civilization, went positive action to control insect-borne diseases and provide cures for them. Since insect-carried diseases are so important in medicine, even a brief account of these measures reads like a short history of medical science.

One role that insects have always played in history is that of depredators. Often this negative role is an important one. Old India hands used to say that there could be no written history in the East because the white ants had eaten all the archives. The Reverend S. Langdon, whom we have already met in his missionary station in Ceylon, recalls ruefully: "I had some fine old works of reference in one of my bookcases, which I had not occasion to use for some days. On going to it yesterday for a volume I wanted, I found that the termites had taken a fancy to these old books, and riddled them through and through, reducing some of them to powder."

The speed with which termites will consume books of paper, parchment, palm leaves, or other substances accounts for the paradox that, although the ancient civilizations of India and Ceylon were brilliant indeed, we know virtually nothing about them or about the art objects they produced in perishable materials—or even in soft stone. All we have left is substantial stone remains, coins, and other metal objects; these the termites could not eat, or have not invariably eaten, though they will chew even concrete on occasion.

All we know about ancient history is what these voracious insects have disdained to eat. The kings of ancient India, for example, used to keep secretaries who wrote down everything that the monarch said, and maintained a complete record of everything he did. But not a single one of these "conversation books" has come down to us. The only ancient civilizations we know anything about are those situated off the termite beat—places like Egypt and parts of Nubia, and Europe. Assyria and Babylon had clay books—nothing else has survived. European civilization has consequently always predominated in culture. It is quite possible, for example, that there were poets as brilliant as Geoffrey Chaucer in Ghana in the fourteenth century, but all we have of ancient African literature is a few folk rhymes and legends that have been handed down orally.

Termites did not confine themselves to eating books, paper, and wood. They ate once a complete English line-of-battle ship, the *Albion*. "Having boarded her," writes the American entomologist Frank Cowan, "in spite of the efforts of her commander and her gallant crew, they got possession of her, and handled her so roughly that, when brought into port, being no longer fit for service, she was obliged to be broken up."

The depredations carried out by insects are more keenly felt, however, when they are directed against food supplies rather than against material possessions. Even if your house falls down because of termites, or your chair collapses under you, or a chest of gold coin in the upper story of a fort crashes down to the basement because all the floors have been eaten by the white ants, you can always camp out under a tree—so long as you have something to eat.

The first serious attempt to colonize Greenland was turned back by a moth. The Greenlanders could not grow grain—Greenland lies too far north—and they were dependent on the dairy products and meat produced from their pastures. By the fifteenth century the climate in Greenland had become progressively less humid, and the pastures were attacked by enormous crowds of the larvae of a moth called *Agrotis noctua occulta,* a relative of the "army worm," which does so much damage to cereals in the United States. Because of this and other factors, such as a general change in climate, to colder, the whole colony collapsed—and not merely collapsed but disappeared with hardly a trace. Nothing is heard of the Greenlanders, or Greenland, till the sixteenth century, when explorers such as Sir Martin Frobisher found the whole island deserted of its former Norse inhabitants.

Many early American settlers have also had trouble with insects. In Utah, in 1847, the Mormons, or Latter-Day Saints, saw their first crop destroyed by myriads of black crickets—the crop-devouring Mormon cricket, or *Anabrus simplex.* In 1848, the crickets were back again, in even greater numbers. The settlers did not despair—

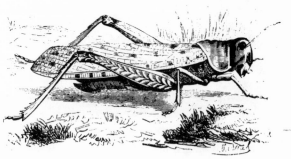

The locust has such powerful wings that it can fly seventy miles a day—and thus spread its damage over very large areas.

they prayed. Flocks of sea gulls arrived and ate up the crickets. The Latter-Day Saints erected a monument to the sea gulls in Salt Lake City, at a cost of some $40,000, which can still be seen.

The fate of these two communities, that of the Greenlanders and the Mormons, has been repeated again and again throughout history. Sudden swarms of innumerable insects appear, destroying a plentiful harvest and threatening the community with famine.

The best-known insect depredator is the locust, or *Locusta migratoria*. The locust is a sort of Jekyll and Hyde. At one phase of its life it is a large, harmless grasshopper. Then the population of these young hoppers builds up, till the locusts relieve the pressure of overcrowding by passing into the second phase of their existence— swarming, or migration. At this point the locust changes color, grows larger, and has great wings that carry it at a speed of seventy miles a day for several days and nights to new areas of infestation. Instead of a harmless ground grasshopper, it has now become an airborne invader, leaving destruction and hunger wherever it stops.

Locusts have been encountered out in the Atlantic at least twelve hundred miles from land. West African locusts were picked up in the streets of London in 1869. The numbers in which locusts arrive and the noise made by their wings are both frightening. One swarm,

Locusts darken an African sky, and man flees before them—as he has fled from this vast depredator since biblical times.

which attacked Kenya in 1928, flew in a column sixty miles in length and three miles in width. It has been estimated that a swarm of this size contains about 50 billion locusts. The prophet Joel described the incredible appetite of these uninvited guests. They tear off the bark of trees, leaving the white branches showing through. They kill the vines, the fruit trees, and the cereal crops. In fact, said Joel, they advance like a devouring fire, leaving the countryside behind them a wilderness.

The despair that struck at the hearts of a population menaced by locusts is generally described in the account of the Eighth Plague in Genesis.

> When it was morning, the east wind brought the locusts. And the locusts went up over all the land of Egypt, and rested in all the coasts of Egypt: very grievous were they. . . . For they covered the face of the whole earth, so that the land was darkened; and they did eat every herb of the land, and all the fruit of the trees . . . and there remained not any green thing in the trees, or in the herbs of the fields.

Whatever Pharoah felt, this must have been *the* plague, so far as the Egyptian people were concerned. After the locust invasion, there would certainly not be food for both the Egyptians and the children of Israel.

Like other unwelcome guests, locusts are a bit insensitive to the unpopularity they generate. Even after they have devastated a country, so that there is nothing left, they will return again the next year. The locust can deposit an incredible number of eggs. In Cyprus, in 1881, by the close of the locust season, one billion egg cases, each containing a considerable number of eggs, were collected, at an estimated weight of 1,300 tons. Yet two years later, a fresh swarm deposited 5,076 billion egg cases.

Another unpleasant consequence of a visitation by locusts has yet to be mentioned. Disease might break out, spread by the flies bred in the rotting mass of dead insects. In 1690 locusts arrived in Poland and Lithuania in three great swarms. An eyewitness called Abbé Ussaris writes, "They were to be found lying in certain places where they had died, piled one on another in heaps of four feet in height. The rains killed these insects, they infected the air, and the cattle, which ate them in the grass, died immediately."

The consequences of a locust invasion were famine and probably epidemics as well. The consequences of locust years would be the decimation—or perhaps the extermination—of isolated communities. Cannibalism would set in—as it did even in Egypt during particularly bad famines. Whole peoples would set off in search of food elsewhere. They would fight with and probably enslave their neighbors. The whole of African history is shot through with these mysterious folk movements, of which the famous *mfekane,* which finally brought the confrontation of white and Bantu in South Africa, is merely the best known. How many movements owe their unexplained origin to a locust famine?

Insect famines have destroyed many communities in history, but they have also strengthened the remainder and acted as a great spur to progress. Those who die in a famine are usually the weakest. This decimation may leave a physically stronger population than existed before—more fertile and able to expand its numbers. Depredations by insects also fall most heavily on a completely agricultural community. In Algeria, in 1866, the locust plague reduced the population by an estimated 5 percent. Consequently, insects have given communities a great impetus to improve themselves, particularly to change from an agricultural to an industrial economy. And this is why, indirectly, agricultural pests have helped to create the modern world of today, and why, in addition to a monument to the sea gulls that ate the Mormon crickets, you can see another monument in America commemorating that great American pest, the

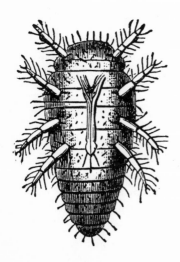

Phylloxera, a variety of American plant louse, so ravaged European vineyards during the nineteenth century that it helped bring about the fall of Napoleon III.

boll weevil. But for the boll weevil, which attacks cotton, the Southern economy might still be largely agricultural. Instead, by helping to make cotton growing unprofitable, the weevil encouraged Southerners to turn increasingly to industry and the raising of other, diverse crops. As the citizens of Enterprise, Alabama, said on their monument to the weevil, it has been "the herald of prosperity."

One last crop devourer might be mentioned, the plant louse *Phylloxera vasatrix*. During 1860–1880 it ravaged the European vineyards so badly that they all had to be replanted with American vinestocks, which were immune to it. Even nowadays American vinestocks are used to grow European grapes as a preventative. So, but for the American vines—which gave America its first name, "Vinland"—there would be no European wines.

2

THE STORY OF SILK

The fact that I am writing this book on an IBM typewriter, and not with a quill pen dipped in homemade ink, is almost wholly due to a nondescript-looking moth called *Bombyx mori*, otherwise known as the silkworm. This tiny moth has had more effect on history than all the rest of the insect world put together.

Bombyx mori is only one of a number of silk-producing moths, most of which live in hot countries. Among them may be noticed the trusser moth, a wild insect whose caterpillars feed on the oaks and other trees in the jungles of central and southern India. The people who live in these jungles collect fresh cocoons from the trees, allow the moths to emerge from them, collect their eggs, and hang them in the trees they have planted in a garden. They watch over the caterpillars as they come out to feed, driving off the rats and bats that would otherwise eat them, keeping up a round-the-clock vigil. At the same time they perform strange rites to ensure the success of the cocoon spinning, which the larvae undertake when they have performed their final molt, or skin casting. Eventually, they kill the embryo moths, still in their cocoons, and plunge the cocoons into boiling water. They are then able to unwind threads of wild silk.

This wild silk is bright and exquisitely soft, but it lacks the strength of domesticated silkworm silk. It is also an arduous and dangerous

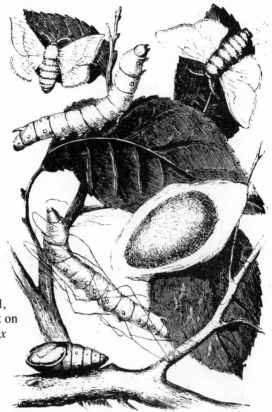

Of the estimated 700,000 insect species in the world, none has had more impact on man's history than *Bombyx mori,* the silkworm. Illustration shows larva, pupa, cocoon, and adult-moth stages of this fabulous insect.

task to collect the cocoons for wild silk. While you are collecting them, a tiger may decide to collect *you.* Altogether it is much more convenient to own domesticated silkworms, which breed, lay their eggs, produce their larvae, and spin their cocoons in captivity. The silkworm eggs can be stored, and by raising the temperature, you can hatch them out pretty well when you like, thus ensuring several crops of larvae, which will spin their cocoons, all ready to be made into domesticated (cultivated) silk.

These silkworms are a little like those other curiosities of nature to be found in China, Père David's deer and the gingko tree. Though they are really wild creatures, they are no longer to be found in the wild but only in zoos and wildlife reserves maintained by man. The *Bombyx mori* have now been living in captivity for thousands of years. They are completely dependent on their human owners, and

According to legend, silkworm cultivation was discovered by Chinese empress Si Ling-chi, who admired the caterpillar's shiny cocoon. Her name gives us the word "silk."

could not fend for themselves in the wild. Careful handling ensures that they survive even though all but a few moths are killed before they emerge from their cocoons. Silk is a material that has become indispensable to man, even now that there are many near substitutes in man-made fibers.

Among the many varieties of silkworms, some of them have a very long pedigree, going back to the legendary start of silkworm cultivation, which was undertaken by the Chinese Empress Si Ling-chi. This lady, the wife of the Emperior Huang-ti, lived 2,600 years before Christ. One day, while walking in the garden of the palace, her attention was drawn to a caterpillar spinning a cocoon with beautiful shiny threads. At the sight of this glistening pale gold being spun into the cocoon, the empress sighed in admiration. How wonderful it would be if she could have a dress made of such material. She asked her husband for a grove of the mulberry trees upon which the caterpillars were feeding and on which the larvae were spinning their cocoons. Huang-ti agreed. The trees in this first silkworm plantation were the Chinese white mulberry, or *Morus alba*. It is from the first part of this name that we get the second word in the silkworm's name—*Bombyx* is the Greek word for "silkworm," and *mori* is the Latin for "mulberry tree." Si Ling-chi (from whose name we get the word "silk") was successful in raising a crop of silkworms. She unraveled the cocoons they produced into long threads, twisting several of them together to make one thicker thread. With the thread so obtained she wove a bolt of silk, which she promptly turned into the first silk dress.

What did it look like? We get a hint in the writings of a third-century monk called Dionysius Perigates, who wrote of the Chinese, or Seres, as he called them (a name which came from the Latin word for silk): "The Seres make precious figured garments resembling in color the flowers of the field, and rivaling in thinness the work of spiders."

Although the story of Si Ling-chi and the silkworms may sound improbable, it could well be true. If an ordinary woman in China had wanted to do something so unusual as starting a new industry, she would have been prevented by the dead weight of male conservatism. If, however, the woman was an empress, people would probably fall over one another to obey her slightest whim. In Old China the ladies of the palace could usually get what they wanted from the emperor.

One wicked concubine, Yang Kuei Fei, the consort of the T'ang Emperor Hsuang Tung, was so extravagant that she could be amused only by the sound of long bolts of silk being torn up. This cruel waste horrified the thrifty Chinese so much that they rose in revolt.

Whether Si Ling-chi ever cultivated silkworms or not, all the other empresses did. There was a special part of the palace garden set apart for growing mulberry trees. Here, on set occasions, the empress would go with her ladies and gather the leaves from three branches of a mulberry tree, which her attendants pulled down till they were within her reach. She used these leaves to feed the royal silkworms, thus setting an example to the nation, for silkworm cultivation had now spread to the whole of China.

There can be no doubt that *Bombyx mori* is a native of China. Nonetheless, China was never the only country that produced silk. India had begun to send out caravans laden with silk at a very early date. The carpets and dyed stuffs of ancient Babylon were woven with gold thread and silk. Aristotle gives a garbled but recognizable account of the life history of the silkworm. He adds that silkworm cultivation was first practiced on the island of Cos, in Greece, by Pamphyla, daughter of Plateos. Did Aristotle know about tusser silk? More likely he had encountered silk raised from the Syrian silkworm, the *Pachypasa alus*. The Chinese silkworm was not introduced into Europe until A.D. 552 (see later in this chapter). Nonetheless, China became so associated with silk that all the old names of the country, Seres, T'sin, Sinem, and Sereca, all mean just one thing: "land of silk."

Silk is a unique substance, a material for which no substitute has ever been found. I never owned anything made of silk until I bought a silk handkerchief I needed for a special purpose. I soon found out that this silk handkerchief was quite different from my cotton or linen ones. When I had a cold, it absorbed moisture much more quickly; it also felt warmer and was much softer.

Down through the ages, man has praised the fineness of silk, its strength and luster. An ounce of natural silk will stretch a hundred thousand yards in length. It is strong because it is produced in a continuous double thread, spun by two spinnerets in the silkworm's body. These special qualities of silk made it suitable for very different uses. Because it is highly *hygroscopic*, able to absorb 33 percent of its weight in water without feeling damp to the touch, it was ideal for clothes worn in the very hot Chinese summer, or for those intended for other hot climates. It was also ideal for underwear. Because it was so strong, it could be put to all sorts of uses. Prisoners languishing in cells at the top of medieval castles could escape if a faithful friend would shoot an arrow in through their cell window with a silk thread attached. On hauling up this silk thread, the prisoner would find a silk cord attached to the end of it, and at the end of the cord a silken rope ladder, down which he could climb. Because a silken rope was so strong and would not break, persons of royal blood enjoyed the doubtful privilege of being hanged on a silk rope. Queen Joanna of Naples had her husband hanged on such a rope—though a handsome young man, he had proved a disappointment to her.

But the most important quality of silk was its beautiful shimmering luster, which made it look like a captive moonbeam. Silk thread has an affinity for rich and delicate dyes. Because the threads of silk are so fine, two threads, each dyed a different color, can be placed side by side and thus produce a color blend in which the two separate parts, the threads, are invisible to the naked eye.

"All good things are to be bought by labor," Leonardo da Vinci

In captivity, silkworms must be carefully tended, fed, and kept clean and warm—as this illustration of a nineteenth-century French rearing shed shows.

once remarked. To produce silk, the silkworm raiser had to work very hard. Like any farm animal, the silkworm has to be fed continually; it has to be cleaned up, given fresh litter, and so forth. To show you just how much work goes into rearing silkworms, let me quote in a slightly adapted form the instructions given to the American colonists in Georgia as to how they should rear their silkworms. The author, an Englishman writing in 1754, tells the readers of his book just how the Chinese go about the task.

When the choice of silkworms is made for breeding they lay the males and females together on a sheet of paper, which must be made of the bark of the mulberry tree. These sheets must be spread on mats well covered with straw, and when the moths have been together about twelve hours, the males must be taken away. The eggs which stick together in clumps must be thrown away, and then the sheets of paper hung up to the beam of the room, care being taken not to turn outwards that side on which the eggs are laid, and that nothing made of hemp may come near the silkworms or eggs. When the sheets have hung in this manner for some days, they are taken down and rolled up loosely with the eggs inwards. The time of hatching the eggs is when the mulberry tree is in leaf, for they are hastened or retarded according to the different degrees of heat imparted to them. When they are ready to come out, the eggs swell and their roundness becomes a little pointed. Then they turn an ash-grey color and afterwards appear blackish. Next day, taking out the rolls of eggs on paper and opening them, they find them full of worms, like little black ants. These insects must be very nicely managed, before their first moulting. Every day is to them a year, and has in it four seasons. The morning is Spring, the middle of the day Summer, the evening Autumn, and the night Winter. Being of this delicate nature, everything ought to be removed that might incommode them. The silkworms have a peculiar aversion to: hemp, wet leaves, the smell of broiled fish, burnt hair, musk, smoke, breath smelling of wine, ginger, etc., as also to all great noises, nastinesses, the rays of the sun, the light of a lamp in the night, much cold or heat, and especially a sudden change from one to the other. With respect to food, the leaves should be gathered two or three days beforehand, and kept in clean, airy places, not forgetting for the first three days to give them the tenderest leaves, cut into little shreds with a sharp knife.

After they are hatched they must have 48 meals the first day, the next 30 and the third less, for, if they be overheated you ruin all. Eating often increases their growth, on which the chief profit depends. The critical moment for removing them into a proper apartment to work in, is when they are a bright yellow, then the numerous swarms must be surrounded with mats at a proper distance, which must also cover the top so as to keep out air and light because they love to work in the dark. However, after the third days's work, they take away the mats from one to three o'clock, to let in the sun, but so that the rays may not strike upon these little laborers. In seven days, the cocoons being finished, they are gathered and laid in heaps till they have time to wind off the silk, setting apart those designed for breeding. To kill the moths in the cocoons, fill great earthen vessels with cocoons in layers of 10 lbs. each, throwing between every layer four ounces of salt, covering them with large, dry leaves, like those of the water lily, then stopping the mouth of the vessel very close. The cocoons must not be put into the cauldron of boiling water till you are ready to unwind the thread, for if they soak long, they will hurt the silk. The silk is wound off by the Chinese with very simple instruments. Five men may wind off 30 lbs. of cocoons in a day and suply two others with 10 lbs. of silk.

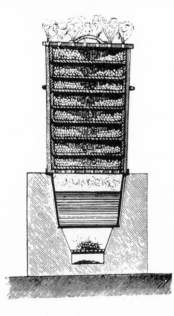

Before the cocoons of silk are unwound, the moths inside them must be killed, which is here done with steam.

Whatever happened to the Georgian silk industry? The sponsors of the colony were very enthusiastic about raising silkworms in the colonies, and the Reverend Samuel Pullein wrote a book to encourage the settlers. But the work was exacting and difficult, unsuited to a raw, unsettled land. The French and Italian instructors sent over to

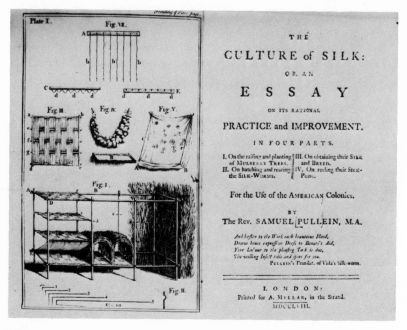

Philanthropists in England founded Georgia in hopes of creating a silk-raising industry, and in 1758 a book was published specifically for the colonists' instruction.

teach the colonists the craft were afraid to let out the secret of their own skill, for fear they would then be out of a job. Some samples of silk were sent to England, where they were very favorably received. But the Georgians themselves found that they simply could not carry out the work of silk farming and earn a living at the same time. Ultimately they took to growing cotton instead.

In fact, for a country to become successful in silk rearing, it had to have a particular kind of social structure. There had to be many peasants whose wives and daughters could do all the work of silk rearing (apart from cultivating the mulberry). China was just the kind of country that enjoyed these advantages, and in addition, it had a centralized bureaucracy, working under a single ruler, the emperor, who could direct a national industrial effort.

From time to time, it is true, emperors did get a little worried about all this agricultural effort being put into a product that you could not eat. Some emperors, including even the husband of Yang Kuei Fei, closed down some silk factories as a protest against all this useless

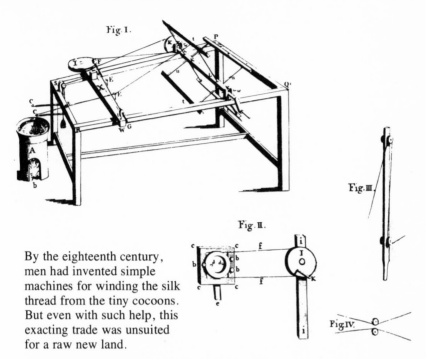

Fig. I.

Fig. II.

Fig. III.

Fig IV.

By the eighteenth century, men had invented simple machines for winding the silk thread from the tiny cocoons. But even with such help, this exacting trade was unsuited for a raw new land.

luxury. The factories always opened again, however, because the Chinese had now discovered that they could pay for all their imports just in silk. By 2000 B.C. silk had become the principal Chinese export. Yu, minister of agriculture of the Emperor Yao, developed irrigation and mulberry culture. Provincial and imperial edicts ordered the planting of the mulberry tree and forbade its destruction. In China, as travelers noticed, even beggars wore silk, and so too, did the Buddhist priesthood, though one of the tenets of Buddhism forbids the taking of animal life.

But the real market for Chinese silk lay outside the Chinese empire. At the time of Alexander the Great, silk was worth its weight in gold in Greece. The Chinese made great efforts to maintain their monopoly. They decreed the death penalty for anyone found exporting the silkworm. For three thousand years their monopoly was complete, and China grew rich on her export trade. The demand for silk from the Roman world was insatiable. Julius Caesar covered the whole of the Colosseum in Rome with an awning, or *velarium*, of oriental silk. Patrician Romans flaunted their silk cloaks, till the Roman Senate felt obliged, reluctantly, to prohibit men from wearing silk at all. Everybody obeyed this edict, except, of course, the emperor, who was a law unto himself. But if the Romans had abandoned silk garments, their wives had not.

Pliny, a Roman writer on natural history, complained that the effect of the almost transparent silk material worn by Roman ladies was such that "a married lady shines through her dress in public, and women's robes denude them." Was his observation correct, or was he merely indulging in the traditional male license to criticize women's clothes? We do not know, because apparently no transparent silk dress material has survived from ancient Rome, not even in the dry climate of Egypt. There was, however, one aspect of these silk dresses which really did shock every Roman husband—their price. The Emperor Aurelian refused his Empress Severina a silk dress that she wanted, telling her that silk was far too dear. In Rome it cost twelve ounces of gold per pound.

The fact that silk had come down in price since the days of Alexander the Great was probably due to the development of two main silk routes and their offshoots between China and the Roman Empire. There was also a sea route, via India, but it never became important for the export of silk to Rome.

All the Roman silk was still coming from China. When the Emperor Heliogabalus treated himself to an all-silk robe, called a *holosericum*, every thread of it had been unwound by the patient fingers of Chinese women. It had all come along the Great Silk Route, a caravan track that began in the Chinese province of Sinkiang. There Bactrian camels were loaded up with bales of the silk, made up in the regulation width of nineteen inches, and inscribed in Chinese and in Brahmi (an ancient alphabet of India) with the place of production, size, and weight of the bale. The silk paid toll leaving China, and it was going to pay a lot more tolls before it reached Rome—which was why Severina could not have her silk dress. The camels made their way through Jade Gate across the desert into Turkestan, where the crossing of the Oxus River had to be made. They met caravans traveling in the opposite direction, carrying jade. Although the Chinese are so fond of this stone, none of it has ever been found in China; it all had to be imported with the money brought in by the silk. Once across the Oxus, the silk was carried into Afghanistan, moving through Bamian, under the gaze of enormous Buddhas carved in the cliffs, till it got to Kabul. From here there was an offshoot of the Great Silk Route that struck south to India. The main route, however, led on into Baluchistan, passed into Persia, and moved via Meshed and Nishapur to Teheran.

In Persia the clever Persian weavers unraveled some of the plain Chinese silk, dyed it, and rewove it into beautiful material. Persian silks continued to reach Europe right into the Middle Ages. Saint Cuthbert of Durham (at whose tomb George Washington's forebear, Prior Washington, must often have knelt in prayer) was actually laid to rest in a Persian or Arabic silk wrapping bearing a Muslim inscription saying that there is no god but Allah and Muhammad is his prophet. From Teheran the silk merchants proceeded through Hamadan to Iraq. There the route led from Baghdad, via Rutba, through the Syrian desert to Damascus, then northward up the Levantine coast to Adana. The weary travelers now moved north past Adana through the Cilician Gates of the Taurus across the flat Turkish plain to Constantinople (now Istanbul), the second capital of the Roman Empire.

Because the Great Silk Route was, in effect, the only road between East and West, anyone who wanted to go from, say, China to India, had to take that route. Many of these travelers were literate people trained to observe, such as the most famous of all Chinese travelers, a monk called Hiuan-tsang, who, in A.D. 629, set off from China to make a complete tour of all the countries of the Indian subcontinent in order to collect Buddhist texts. These travelers have described the difficulties and dangers of the route for us. They stood in constant danger of attack from the marauding tribes of Central Asia. It was usual to hire guards to protect the caravan, and move in large parties, even if it meant waiting at an oasis for some time while more merchants arrived. Parties were sometimes buried by sandstorms in the desert, or arrived at a desert water hole to find it had dried up. Another adventurous traveler, an Italian merchant named Marco Polo (1254–1324) writes feelingly about how evil spirits imitate the noise made by the caravaneers' musical instruments, so that stragglers are led off the right path and die of thirst in the desert. Other dangers included the incredibly narrow paths through the mountains. One false step on the part of your mount could plunge you off the icy path into a torrent running at the bottom of the precipice far below you.

Wherever the Great Silk Route or its offshoots passed, the

inhabitants of the region became enriched by providing food and water for the travelers and their horses and camels and by taking tolls. Some of these dwellers in Central Asia were not the sort of people we would expect to find there today. The inhabitants of the Turfan oasis in the central Gobi desert, for example, were tall, fair-haired, blue-eyed Indo-Europeans. Archaelogists following the trail of civilizations left on the Great Silk Route have discovered the most amazing cultural riches in lands that are now virtually deserts. Sir Aurel Stein, a Jewish archaeologist who had left his native Austria to take service with the British government in India, discovered Buddhist silk banners and paintings, and also the oldest book printed in the world, dated A.D. 684. Sven Hedin, a Swede, found some of the oldest Chinese writings ever to be unearthed, and Von Le Cocq, a German of French Huguenot descent, unearthed caves full of sculptured and painted Buddhas, some of the best of which went to Berlin, to be destroyed by bombing during World War II.

Where the silk route passed through the Chinese or the Roman Empire, it was relatively safe to travel on. But both emperors tried to push their frontiers farther and farther out, to control more and more of the silk route with its lucrative tolls. From the Roman side, this policy proved a big mistake, for many of the independent kingdoms they destroyed would have acted as a barrier between them and the hurricane that was about to break on them from the east. In A.D. 115, the Roman Emperor Trajan made Mesopotamia a Roman province. People in Rome expected him to go on to invade China, or at the very least push Roman conquests up to the very edge of the Chinese Empire.

On the Chinese side a general called Pan Ch'ao extended Chinese power into Central Asia, where there were a number of formidable nomadic empires, such as those of the Mongols, the Tartars, and the Huns. In A.D. 97, the Chinese ambassador, Kan Ying, was sent by the Chinese emperor to represent his interests in the Roman Empire at Antioch. East had met West at last, and it had taken the threads of the silkworm to pull them together.

On the western side a Macedonian merchant called Maes Titianus began a systematic exploration of the Great Silk Route. His agents reached Loh-yang, capital of the Chinese Han empire, and among other information, brought him back the news that silk did not grow on trees. The Chinese had pretended that it grew on branches in tufts, like cotton, so as to safeguard their secret. Titianus now learned that the precious substance was produced by a caterpillar.

The Chinese policy of policing the Great Silk Route was to have consequences of incalculable historical importance for the West. What had begun as a police action turned into a war that lasted for centuries. The Chinese would not pull back—that would have been to lose face. Yet they found the resources of even their vast empire strained past the breaking point in fighting the mobile Hun archers. So formidable were the Huns that they thought nothing of riding out of camp before dawn, destroying a Chinese army fifty miles away with their arrows, and getting back before nightfall. The Chinese had to fall back on fortifications, and new weapons such as a repeating crossbow that fired poisonous bolts. They also used allies, such as the Indo-Scyths, who fought in the same way as the Huns.

After three hundred years of continual warfare, the Huns decided they would give up the struggle and turn their horses' heads westward. Just what prompted this decision we do not know. The Chinese had inflicted enormous defeats on the Huns, destroying the western half of their nation. But the Huns had retaliated, overrunning Chinese territory and even capturing the emperor's capital of Ch'ang-an. Possibly the Hunnish decision to go west was made because of rumors that traveled along the Great Silk Route about the symptoms of decline that had begun to appear in the Roman Empire. The Huns were not mechanically minded and found it difficult to besiege walled towns, such as the Chinese held. They knew, though, that between them and the Roman Empire lay whole nations of barbarians—the Alans, the Goths, and the Vandals, who lived a simple nomadic life like themselves. Why not attack this easy prey? Distance was no object to the Huns, and they would as soon lift the scalps of Western Barbarians as Chinese or Indo-Scyths. In A.D. 372 their chief, Balamir, began to lead them westward.

They swept the barbarians before them like chaff, and these peoples, finding that there were only two choices for them—to be crushed by the Huns, or to force their way into the Roman Empire—chose the latter course. One by one the barbarians I have mentioned, and in addition the Ostrogoths and Visigoths, entered the Roman frontiers, first as tolerated fugitives, then as allies, then as invaders, and finally as conquerors.

What made the barbarians irresistible in battle with the Romans was that they had picked up from the Huns the habit of riding in iron stirrups, which was apparently a Chinese invention. This may seem rather a small invention to overturn an empire, but the barbarians, using their stirrups which gave them a firm, braced seat in the saddle, were able to do something no Roman cavalry had ever dared to do—charge right into infantry ranks. Then the Huns themselves arrived on the heels of the barbarians. They invaded the empire in 453, were turned back from Rome by the personal intervention of Pope Leo, and defeated at Adrianople. It made no difference in the end. The Roman Empire collapsed under the multiple shocks imposed on it.

An obscure group of barbarians, living in the depths of the German forests, who had made many attempts to establish themselves in the British Isles, now saw the way open to them at last, for the Roman garrisons in Britain were being withdrawn to meet pressures elsewhere in the empire. The Angles, Saxons, and Jutes streamed across the North Sea in their longboats and invaded Britain in force, creating the English kingdoms and, incidentally, the English language.

Even the fall of the Roman Empire did not interrupt trade along the Great Silk Route completely. The caravans carried not just bales of silk, but ideas and inventions as well. Eastward went the vine and the technique for making wine, still written in Chinese by two ideographs that mean "water of the West." Nestorian monks from Syria moved eastward into Persia. Eventually they reached China, where the emperor made a decree in their favor. But meanwhile two Nestorians had succeeded in destroying the Chinese monopoly of silk. According to tradition, about A.D. 300 a Chinese princess who left China to marry the king of Khotan (on the frontiers of India) carried with her silkworm eggs as a secret present for her husband. She had persuaded her silk moths to lay the eggs on the flowers that decorated her hair, probably made from "rice paper"—that is, the pith of the *Fatsia papyfera* tree. If the princess had worn any other kind of flower, the silkworms would simply have died before she reached her journey's end. So the silkworm had now moved outside of China, but not very far, only to a country on the borders of India.

The Emperor Justinian of Byzantium may perhaps have heard of these silkworms, which had traveled to Khotan, and decided that perhaps they could travel a little farther. He persuaded two Nestorian monks, who lived in Persia but who had traveled in China, to go back there, collect some silkworm eggs, and conceal them in the hollow joints of a bamboo cane. In A.D. 552 the monks did what the emperor had commanded them, walking unperceived through the frontier guards carrying their bamboo staffs. Justinian was delighted. He had bypassed the stranglehold exercised on the silk route by his great rival, the king of Persia. He now set up a special factory in Constantinople, bringing weavers from Tyre to work the silk under the direction of the monks. When this story leaked out, the Nestorians who remained in China must have become suddenly unpopular. Hitherto they had been very influential; they had even given an alphabet to the Manchus, the nomadic warrior tribesmen who were eventually to invade China and set up the last Chinese dynasty. Now they disappeared mysteriously.

The loss of the monopoly of the secret of sericulture (which soon spread outside of Byzantium) did not mean an end to East-West trade. More and more ideas began to reach Europe from China. They include the use of the artist's brush, which the Chinese had invented for writing their characters, the invention of printing from wood blocks (this was how the earliest printed book, found by Sir Aurel

In nineteenth-century France, women workers in a silk mill ravel the filaments from the cocoons onto reels, under the watchful eye of a forelady.

Stein, had been printed), paper (a Chinese invention that had been discovered through observation of the wasp making its nest), gunpowder, and firearms. Marco Polo alone brought back three Chinese ideas: growing rice in irrigated fields; making macaroni out of flour paste; and using paper money.

By the sixteenth century, silk raising had become established in Italy and France. It was tried in other parts of Europe where the white mulberry would grow, but without much success. A lot of hands were needed to unwind the silk cocoons, so the Italians devised clever machines to speed up the processes involved. The town of Bologna was among the first to have "throwing mills" or special workshops, where machines were used to twist up the silk fibers for the weaver.

One country that was determined to acquire its own silk industry was England. Hand weavers of silk in England were mostly Protestant Frenchmen, who had fled to England to escape Louis XIV's attempt to force them to become Catholic against their will. They settled down at such places as Spitalfields and Soho in London, where you will still hear many people exchanging greetings in French even today, for it has become a sort of foreign quarter of London. These foreign workers produced beautiful silks.

But early in the eighteenth century, a man called Thomas Lombe learned Italian, went to Italy, and pretended he was out of work and wanted a job. He got taken on at one of the throwing mills, where he secretly made drawings of the new machines. Then he returned quickly to England, fearing for his life if the Italian sericulturists caught him, and set up his own factory at Derby. The factory, built in 1718, stood on the River Derwent. The throwing mill was completely mechanized, and all the machines were driven by waterpower. Just how modern and sophisticated Lombe's machinery was can be seen from the fact that his principal silk-throwing mechanism had 26,586 wheels, which worked 73,726 yards of organzine silk thread with every revolution of the waterwheel. Organzine silk thread was the strongest and best kind of silk thread, used to make the warp in fine silks. Hitherto the silk produced by English silk throwers had only been strong enough for the weft. Since the waterwheel revolved three times a minute, the mill could produce 318,504,960 yards of organzine daily. Lombe's factory was five stories high and an eighth of a mile in length.

Thomas, or Sir Thomas, as he later became, had been successful in accomplishing what he had set out to do, building up an English silk-throwing industry. By 1765 there were seven organzine silk mills in England, designed on Lombe's principles and employing hundreds of workmen. All these workmen worked inside the factory. This was something new. Up to then if you wanted any spinning done, you took out the raw material and left it in bundles with cottagers' wives and daughters, whom you paid a low wage to spin it at home. The women operating this cottage industry recouped themselves for their low wages by keeping back some of the raw material you had given them—stealing it, not to put too fine a point on it. Silk was too expensive to be handed out in this way, so a new principle of work had been established: work had to be done inside the factory.

But there was one big stumbling block to the English silk industry—the difficulty of providing raw silk. The English climate was all wrong. The white mulberry would grow in England. One was

This white mulberry tree, food of the silkworm, was planted by the monks of Lesnes Abbey in Kent, sometime before 1536, and still thrives today.

planted by the monks of Lesnes Abbey in Kent, which is still growing today. Likewise, Chelsea can boast a white mulberry that was planted by Sir Thomas More, the chancellor of Henry VIII, whom Henry later beheaded. More may have planted it as a pun, because the name of the tree, and his name, are spelled the same way in Latin—*morus*. But the silkworm, a Far Eastern insect, was used to the very hot summers of China. To keep it warm enough even in sunny France, the rooms where it lived had to be heated with big stoves, which was very expensive.

In my lifetime there has been only one silk farm in England, that run by Lady Zoë Hart Dyke, at Lullingstone Castle, in Kent. Even this one farm was stretched to capacity to provide the twenty pounds of raw silk required to make the coronation robes of Queen Elizabeth, the Queen Mother, and the little Princesses Elizabeth and Margaret. That was in 1937. Some time after providing silk for Queen Elizabeth's coronation robes in 1953, the farm moved to Ayot Saint Lawrence.

England had hoped at one time to get its raw silk from Georgia. Raw silk from Georgia and South Carolina was exempted from duty in 1749, and a bounty was offered for its production. But as we have seen, silk cultivation was ill suited to life in the wilderness. For one thing, raising silkworms has always been women's work, and there were just not enough women among the colonists to tend the silkworms. In colonial America there was always an imbalance between men and women, with far too few of the latter.

If silk could not be raised in Georgia, what could? The answer, we have already seen, was cotton. Georgia and the rest of the South began to grow cotton. In England, the factory system, which had been devised for silk, was simply shifted to cotton processing. Cotton had been spun and woven in England for quite a long time, but it had come in small quantities from the Greek islands and been handed out by middlemen called "broggers" to the home spinners and weavers of England, who worked in their own cottages. These workers worked hard, but they were independent. They could refuse to accept a job or stop work whenever they liked to cultivate their gardens or even go off to a village merrymaking. The industrial entrepreneurs of England had never liked employing the cottage-industry workers. It was impossible to control the quality of their work, or to ensure that it got done on time. And there was always waste through pilfering.

Altogether, the industrialists thought, it would be much better for their purposes if they could confine the newly developing English cotton industry to factories of the Thomas Lombe type. Everything would be done by machinery. Because the workers merely had to tend a machine that was being driven by waterpower, and did not have to work it by pressing a foot treadle, they could work harder. Because the industrialists wanted to get the most out of their expensive mills, they could keep them going day and night, and workers, whether they were on the day or the night shift, would work very long hours. Women and children would work for less than grown men, so the employers preferred them as a working force. If they were late at work or could not keep up with the ever-increasing tempo of the machines, they were fined by having their pay docked.

Man had hitherto worked at his own pace, and his pace had been a leisurely one. It might have taken a peasant two hours to load up a packhorse. Man's only clock had been the sun. Now the factory worker began to work faster, and ever faster. His actions took on the frenetic pace we associate with modern man. He could no longer go out to relax by digging his garden—that would mean the imposition of a fine—and in any case his garden and his cottage now belonged to someone else. Instead, he and his family were living in a noisome slum owned by the industrialist, who built shoddy and cramped housing accommodations for his workers. And all around the factories the green and pleasant land of England had disappeared in a rash of urban development, through which ran polluted rivers.

It was not long before the factory owners felt that the wheels could be made to move a little faster than waterwheels could drive them. As usually happens when there is a demand for a new invention, an inventor stepped in and filled the requirement. A Scot named James Watt, in partnership with an Englishman called Matthew Boulton, developed the first machine that produced its own power, the improved steam engine. The Age of Technology had begun.

The *Bombyx mori* still had one little surprise in store for mankind. Around A.D. 200, Japanese commercial spies had penetrated the secrets of Chinese sericulture. They had obtained silkworms and their eggs, and they had persuaded four Chinese girls who knew how to manage them to leave China for Japan. "Kidnapped" might be a better word than "persuaded." These girls taught the women of Japan silkworm breeding, silk weaving, and embroidery in silk. The

Japanese produced some of the most wonderful silk textiles in the world. Some of them can still be seen, because many precious things were laid away in a special safe deposit, the Shoso In, by the Empress Komyo of Japan in A.D. 823.

The Japanese, who learned the secret of silk culture from the Chinese about A.D. 200, still provide some of the world's most beautiful silk textiles.

When Commodore Perry arrived in his "black ships," as the Japanese called them, off Japan in 1853 and ordered the Japanese government to open trade relations with the West, the Japanese were a bit puzzled as to what they *could* export. They offered Perry samples of all the things they did produce, including toilet paper, which was then unknown in the West. Perry got indignant, thinking the Japanese envoys were trying to make fun of him. But Perry, and everyone else, admired the Japanese silk textiles very much. Japanese silk became the country's most important export.

Forty percent of the farmers of Japan, more than two million households, raised silkworms. One tenth of the arable land of the country, a country that is more mountain and forest than farmland, was planted with mulberry trees. The Japanese farmer's wife and daughters worked long days and nights raising four crops a year of *okaimo sama* (the honorable silkworms). By 1931 the raw silk exported paid for all the cotton, coal, rubber, and automobiles Japan

After Japan was opened to trade with the West, silk became her most important export. Above are shown some of her nineteenth-century silk-weaving patterns.

imported in that year. Next to rice, raw silk had become the most important Japanese product.

Japan's principal export was almost all going to just one country, America. Ninety percent of Japanese silk was now taken by the United States. Japan's economy was tied up with that of America's. In 1929, when shares of U.S. Steel sold for $260, raw silk touched $700 a bale. Then in August, 1932, after the Great Crash of the stock market in America, when U.S. Steel had dropped to $26, raw silk was selling for $153 a bale.

The Japanese have always been good at doing sums. Here was an arithmetic problem that required little skill to solve. Either Japan was going to become a fourth-rate industrial power, selling her only export for about one fourth of what it had been worth before the Great Depression, or she must abandon the way of honest trade and try to bolster up her economy by acquiring wealth and territory overseas. She chose the latter course, thus ultimately drawing the United States into World War II and unleashing on herself the most fearful product of the Age of Technology, the atom bomb.

The development of the silk industry in England had produced the Industrial Revolution, centering in the cotton industry, then gradually extending to all industries. It had begun in England in 1760, and had gradually spread to all the major countries of the world. It is still spreading now. The Industrial Revolution was probably the most important event in history, and it largely had been promoted by a little moth so small that you could cover it with a nickel.

3

SINGING INSECTS

$\int$ome insects, far from spreading disease or making nui-
sances of themselves, have lived very much in harmony with human
beings. People have made pets of them for their musical abilities,
listened entranced to their song, written poems about them, treated
them as one of the family, and housed them in fancy cages, which
were sometimes more costly than the owners could really afford. The
owners of these singing insects usually loved them dearly and often
"put them out to grass" by releasing them. They have wept bitterly at
their deaths, erected special tombs to their memory, and even
worshiped them as sacred beings, very close to the gods.

In some ways it is misleading to talk about "singing insects," for
insects have no vocal organ and consequently no voice. What they do
is they produce music by turning themselves into musical instruments.

Take the short-horned grasshopper, for example. This insect is one
of nature's violinists. On the long legs that it uses for jumping is a
ridge, topped by a row of very tiny knobs. This serrated edge is the
grasshopper's violin bow. By moving its leg, the short-horned
grasshopper rubs the "bow" against the upstanding veins of its closed
forewing. Those veins or ridges act as a violin to the bow, and as the
two rub together, the music is produced.

The long-horned grasshopper and the cricket make music in rather a different way. One of the lower forewings of the cricket has prominent veins or ridges, similar to those of its short-horned relative. The underside of the upper forewing is ridged like a file. This is the bow, and when it is drawn across a sharp-edged vein in the wing below, it produces the shrill music for which the cricket is famous. The space between its folded wings and the back acts as a kind of sound box to intensify the volume of the music, like the belly of a violin. Music is so important to the cricket that it has a "second fiddle" in the other forewing and lower wing, in case anything goes wrong with the first.

The field cricket, which rubs its wing ridges together to produce its chirping song, is one of Nature's most appealing music makers, beloved all over the world.

The long-horned grasshopper only has one "bow," a filelike ridge found on the underside of the left forewing. This is kept folded above the right forewing, the upper surface of which has a sharp edge, like a violin string.

Besides grasshoppers and crickets, there is a third group of prominent insect musicians—the cicadas. The music of the first two groups can be compared to the rubbing of a violin bow across the strings of the instrument, or, perhaps more accurately, to a xylophonist running his wooden hammer along the bars of his instrument. The cicada, on the other hand, can best be compared to the big drummer of the orchestra. It makes its music by vibrating a drumlike membrane. Each male cicada has two of these drums on the underside of its body near the waist. By tensing up the muscles behind the drums, the cicada brings them into play. It contracts the muscles inwardly, then releases them suddenly, an action that throws the drumlike membrane into a state of vibration. Like the big drummer in an orchestra, the cicada can easily drown out all the other performers. One Brazilian species gives out a roar like the whistle of a locomotive, and another member of the family can be heard at a distance of a mile away.

Seeing that insect musicians go to such trouble to produce music, it is obvious that they must hear it. Cicadas can apparently hear without ears, because they do not possess them. They probably "hear" through the vibrations they feel. Crickets and short-horned grasshoppers have their ears in their great leaping legs.

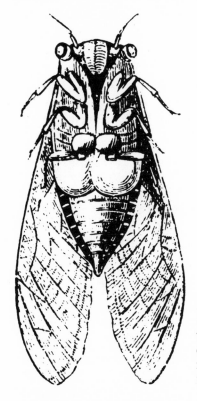

Another insect musician is the cicada, which has two drumlike membranes on the underside of its body and makes sounds by causing these to vibrate.

Insects are attracted by the music made by human musicians. They will approach when a bell is rung or the strings of a musical instrument are plucked. One ancient Greek musician, according to an old legend, was playing in a competition, when he had the chagrin to see one of his lyre strings break. It looked as though he were going to lose the contest, but a friendly cicada settled on the instrument and continued to sound the missing note.

Music making, as you will have noted, is confined to the male insects, and Xenarchus, an ancient Greek poet and woman hater from Rhodes, said that cicadas are very lucky, for their wives never have a word to say. Presumably insects sing to attract females, however, as early as the seventeenth century, an entomologist pointed out that the only singing insects that refrain from singing are the sick ones. All the others keep on singing, all their lives, even when they have no females to attract.

Singing insects also respond readily to weather conditions, the temperature, the time of year, and the time of day. If their sunny-time song is their normal music, then that which they keep for dull days sounds much more doleful.

If you were to take a walk through the middle of a city in Central Africa after sunset, you would see why singing insects have always been popular. Blantyre, in Malawi, where I spent some time, is a very quiet place once the short tropical twilight has pulled down the curtain of the dark. There are no night clubs or cafés because

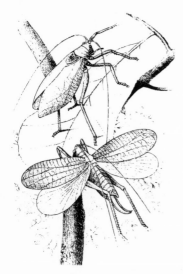

In Central Africa, tree crickets—similar to the species shown here—become a kind of talking thermometer, their chirps coming more and more frequently as the temperature rises.

Malawians are essentially family folk who like to stay home best. In fact, there is nobody with you as you walk down the broad streets, except for the security guards, dressed in old army greatcoats and builders' helmets, who emerge from their doorways for a moment to give you a cheerful smile and a word of greeting. Yet you do not feel lonely because the trees and the ornamental gardens are full of tree crickets. They keep up a loud, continuous chirping, which accompanies you right down the street. Singing insects provide companionship. They are the canaries of the insect world.

The nights in Central Africa are cool at this altitude. The crickets enjoy the warmth, and as the temperature rises, their chirp becomes more frequent. If you have a stopwatch, you can tell the temperature, just by listening to the crickets. One hundred chirps a minute is the equivalent of 63 degrees Fahrenheit. Singing insects are a talking thermometer.

As you walk along the sidewalk, the tree crickets keep up their song until you are quite close; then they stop. They can sense your presence and fall silent. If, as sometimes happens in Blantyre, there is a leopard prowling around among the trash cans a few blocks away, then the tree crickets near to it will stop singing, and you will be warned of its presence in time to turn back. So singing insects are excellent watchdogs.

This is one of the reasons why insects that sing at night are kept as pets in Japan. You might think their song would keep everyone awake. Instead it gives the occupants of the house a sense of security. As long as the cricket is singing, nobody is trying to break in. The singing insect is rather like the old-fashioned watchman who used to walk the streets of London, ringing a bell and shouting, "Four o'clock of the morning and all's well."

The last and most important reason for keeping singing insects is because of their musical ability. Not everyone throughout history has found the music of insects charming. We are told that the legendary Greek hero Hercules was awakened by some cicadas from a siesta on the banks of a river opposite to where the town of Locris stood. He prayed to Zeus that they should cease, and according to the legend they have stopped singing near Locris ever since. Few people in the ancient world would have agreed with Hercules. "As music soothes the mind and dissipates fatigue," says the Greek philosopher Plato, "so the plowman loves and cherishes the cicada for its song." Listening to singing insects is like listening to the sea on the beach, or the wind sighing in the trees. It lulls and soothes the spirit. Here is what an ancient Greek poet wrote about the cicada:

> *We bless thee, cicada,*
> *For that, seated on the treetops,*

Sipping the dew, thou singest royally.
O sweetest of summer prophets!
Honored by mortals.
Thou art cherished by the Muses.
Phoebus himself loves thee
And gave thee thy shrill song.

In ancient Greece the cicada was the symbol of song. Cicadas loved music, Plato tells us, because "as the story goes, before the Muses lived, the cicadas were men on earth, and so loved song and singing that to lose no time from it they left off eating, and so died of that dear delight. But in death they became cicadas, and this boon was granted them by the Muses, lately born, that on earth they should eat no more, but only sing until they died again, and that then they should return to the Muses to tell them, who, amongst mortals, loved and worshipped them most."

Not merely was the cicada the symbol of music, it was also the emblem of the native-born Athenian. Only he could wear the golden bodkin with its head molded into the shape of a cicada, an ornamental hairpin shaped like a dagger, which men used to tie up the hair in a special knot.

The Greeks kept cicadas and other musical insects in small cages made of reeds. The cicadas were caught by means of twigs coated with birdlime, often by the hands of a professional fowler. Pliny, the Roman naturalist, tells us how boys fished for male crickets by catching a fly, tying it to a thread, dusting it, then dropping it into the burrow of the cricket. The cricket seized its prey and was pulled out, becoming a captive rather than letting the fly go.

Although the Greeks knew that only males sing, they kept their insects in pairs, partly so that they could observe their mating behavior, partly because they were kind-hearted folk. Special diets were given to the captive minstrels. Scallions are recorded as being one of the foods. The Greek poetess Anytie wrote a poem about a friend called Myro who kept a grasshopper and a cicada as pets. After their death Myro built a tomb for them, where she sat weeping.

Insects were not nearly such popular pets with the Romans as with the Greeks, but even after the fall of Rome the hobby persisted. On Malta, people still keep cicadas in fish-shaped cages. In Spain crickets and other insects are kept in cages to sing during the celebration of Mass. Spanish insect pets are kept in pairs, in light and airy two-story cages suspended from the ceiling. One singing insect even made a voyage to Brazil. A soldier who sailed with Cabeza de Vaca's expedition, and who was in poor health when the voyage started, took with him a ground cricket in a cage, thinking the insect's song would amuse him. After an obstinate silence all the way out from Cádiz, the cricket suddenly started singing. Could it be that the creature had scented land? The Spaniards looked westward. There,

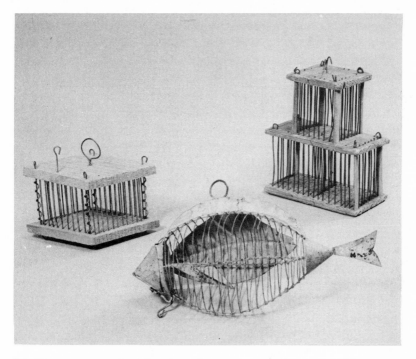

Mediterranean peoples often keep insects as pets. Here are cricket cages from (*left*) Portugal, *(right)* Spain, and *(foreground)* Malta.

within bowshot, the surf was breaking over high rocks. The ship had just time to drop anchor.

Even in North America, far away from Spanish influence, the habit of keeping singing insects persisted for a time in the early American pioneering days. Many German immigrants had kept them as pets as boys. "The youth of Germany," says Frank Cowan, "are extremely fond of field crickets, so much so that there is scarcely a boy to be seen, who has not several boxes made expressly for keeping these insects in. So much delighted are they too with their music that they carry these boxes of crickets into their bedrooms at night and are soothed to sleep with their chirping lullaby."

At least one American lady kept crickets, a Mrs. Taylor. She told her readers in Harper's Magazine how she had obtained several house crickets in the old castle of Caernarvon (in Wales). "These she carried with her in her journeyings to and fro over the Kingdom, for several years, and at last brought them to this country where they were liberated in the snuggest corner of a southern hearth." This lady, knowing that insects are attracted by music, kept one of her pets, called Queen Bess, on the sounding board of her Aeolian harp.

The cult of singing insects was never so strong in America or Europe—not even in ancient Greece—as it was in the Far East. As early in the history of China as the first dynasty, Shang, the Chinese already regarded the cicada as the symbol of eternal life. The larva of this creature lives for seven to thirteen years underground as an ugly white grub. Eventually it tunnels back to the surface and becomes a long-legged nymph. This insect form climbs a tree, splits its skin, then flies off, leaving a cast-off husk. The analogy between the two stages of cicada life, grub and flying insect, and man's life on earth and the hereafter, was irresistible even to the very worldly Chinese. The Emperor Hang Ts'ung (who lived in the eighth century A.D.) said explicitly that the cicada is the emblem of the passage of man from mortal life to a higher stage of being.

The Chinese, then, prized the cicada because its small chirp was a reminder of immortality, because it sang, and because it appeared with great regularity on the arrival of summer. Politeness has always gone a long way with the Chinese. Here was a punctual and polite insect which did not keep anyone waiting for its arrival. They also prized the cicada because it was thrifty—it did not eat any cereals consumed by man but fed, so they thought, entirely on dew.

It is one thing to admire insects and another to persuade them to settle in your home as pets. The Chinese have always shown a genius for keeping pets. They developed the unattractive-looking carp of their ponds into the extremely decorative goldfish. Like goldfish, insects were kept because they were a tiny part of nature that could be accommodated in the cramped conditions of urban Chinese life. No home was too small for a singing insect. In fact you did not even need to have a home. A singing insect could be carried around in a special sleeve cage, where it would provide companionship for its master by giving an occasional friendly chirp. I have one of these cages, made of bone, in rather crude workmanship. This cage belonged to a poor man.

At the other end of the social scale would be the cages owned by the emperor's ladies. "Whenever the autumnal season arrives," says a Chinese historical work of the eighth century, "the ladies of the palace catch crickets in small golden cages. These, with the cricket enclosed in them, they place near their pillows, and during the night hearken to the voices of the insects."

In China, singing insects were kept by professional insect collectors. They bred them, sacrificing all the females, and also added to their stock by catching them. Quite a number of cricket traps have found their way to collections in America and Europe. They are intricately carved out of bamboo or ivory. There is a sliding trapdoor that can be closed, no doubt by a thread, once the insect has entered. Some of these cages are of such rich workmanship that they must have belonged to amateur rather than professional catchers. Insect catchers would walk at random along the outer walls of gardens and old buildings. When the sound of a cricket or some other insect alerted them, they would tempt it toward them by imitating its call and offering it a slice of fruit. Then they would coax it into a hollowed gourd, which was one of the regular receptacles for crickets. Other insects would be driven from their holes by flooding or by the heat from a lighted brazier, or they would be lured out by burning candles.

The captured insects were housed in different ways, according to the season. In summer they were kept in circular pottery jars made from common burned clay, covered with a perforated lid. In these thick-walled jars the crickets could keep cool. They were fed daily on slices of cucumber and lettuce, or other greens, and given fresh water. In the autumn and winter they ate chestnuts and yellow beans, which their owner chewed up for them first so that they could digest them more readily. In south China, ailing insects were given chopped mosquitoes and honey as a tonic. These delicacies were served up to them in dishes of blue-and-white porcelain. They had special miniature water troughs and beds or sleeping boxes of clay.

In winter the crickets moved to a warmer home—usually a hollowed out gourd shaped like a bottle or jar. The best gourd specimens—those for the emperor's cricket stable—were decorated with designs in relief, impressed on the gourd while it was still growing. This technique, which is now lost, was carried out by enclosing the gourd flower in a clay mould with embossed sides. The gourd would then grow and swell out until it had received the impress of the mold. There must have been many of these cages in the imperial collection, and they were looked after by a special official who had to produce whatever singing insects the emperor wanted. Into the bottom of the gourd would be pressed cement of lime and sandy loam as a sort of carpet for the cricket to sit on. At night it would be given cotton padding on which to sleep.

Some of the gourds in which crickets were kept came from a special rare, mottled variety. Young men ruined themselves in buying these objects—or so a rather puritanical Manchu writer of the nineteenth century has complained. To the Western eye, the most decorative aspect of a cricket gourd was its stopper. Stoppers were given perforated flat tops, carved in a gourd-leaf-and-fruit design, and made from the richest materials, such as tortoiseshell, ivory, jade, or walrus ivory stained to look like jade. Often gourds did not stand on a shelf, but hung by means of a bamboo hook from a wooden pole that ran across the room just below the window. This allowed a free circulation of air around the cage, keeping it cool. Other cages took the form of pierced boxes of wood or ivory of various shapes, cages of gold or silver wire or of split bamboo, or porcelain cages with pierced open-work sides. Often these have been brought to America by collectors who did not know what the cages were intended for.

The number of types of crickets that were collected was considerable. There was the mitered cricket, the broad-faced cricket, the besprinkled cricket, the yellowish tree cricket, the black tree cricket, and the spinning damsel, as well as the poetically named golden bell. Different insects gave different notes. The emperor had an orchestra of musicians, each sounding a different note. To enhance the song, the tympanum of good singers was coated with a minute layer of wax. Owners of poor performers asked if they might call with their pets on those who had good singing insects so that their specimens might listen to the song of the star singers and learn to improve.

Unfortunately Chinese interest in crickets did not stop short at singing crickets—it extended to fighting ones. Good fighting crickets were believed to be the reincarnation of great human heroes of the past, and they were called generals and marshals. To the Chinese the cricket fight offered all the attractions that the bullfight has for Spaniards of today, with the added attraction that heavy bets were laid on the contest.

At two localities near Canton, Fa-ti and Cha-pi, not far from Whampoa, regular tournaments still took place between crickets during the 1920's. The fights were held in big mat sheds, outside of which were placards displaying the merits of the combatants. Great sums were staked—as much as $100,000 might be wagered on a single contest.

The owner of the winner was presented with a roast pig, a piece of silk, and a gilded ornament representing a bouquet of flowers. He was escorted back to his house with gongs playing and flags flying, and flowers were scattered before him. The name of a victorious cricket was inscribed on a gourd-shaped ivory tablet, which was hung up in the proud owner's home. When the cricket died, it was buried in a small silver coffin.

Alison Stillwell Cameron, in a delightful passage, describes how her family cricket (a singer, not a fighter) was buried in the middle of the courtyard of her Peking home in such a coffin. The fact that this reminiscence is given not in an autobiography, but in a book about Chinese painting, shows what a powerful attraction insects had as pets.

The fighting insects were placed together in an earthenware jar. Male crickets in captivity often attack one another, so a fight might start naturally. If it did not, then the owners "tickled" the crickets with bamboo fibers or rat hairs held in ivory holders. The fighting crickets were fed on a strengthening diet that included rice, fresh cucumbers, boiled chestnuts, lotus seeds, and mosquitoes, which were occasionally allowed to feed on the cricket's owner before being fed to him.

The Japanese imitated the Chinese in their fancy for singing insects, but not for fighting ones. The insect season began, and still begins, on May 28, at the time of the Festival of the Fudo Temple in Tokyo. The insect sellers, called *mushi uri*, set out their stalls, which are visited by insect fanciers. They buy the insects, which are housed in cages of split bamboo made to look like houses, junks, or even fans. The owners hang the cages under the eaves of their houses or in trees, so that they will break the stillness of the long summer nights with their song. They are fed on fresh eggplant or cucumber daily.

Listening to insect music, which broke the stillness of the long summer nights, was a favorite pastime in Old Japan.

Grasshoppers, green locusts, katydids, cicadas, and crickets are all kept in cages, but cicadas are not nearly so popular as they once were either in Japan or China.

When autumn comes, the little captives are liberated at the Insect Hearing Festival—a rite that has been carried on in Japan since feudal times. In late August or early September, insect lovers gather in temple or shrine precincts, mountainsides, wooded glens, or just their own gardens. In their hands they carry the cages of their insect friends, from which they are soon to part, or even insects bought especially to liberate on this occasion. The cages are placed on the ground, the doors opened, and the insects allowed to depart. There are some moments of silence while the released captives get their bearings, and then their former owners listen in rapture while the insects realize their new freedom and burst into song.

In some ways the story of singing insects is the most important in the whole history of man's adventures with the insect world. It is one of the very few attempts by man to achieve *symbiosis*—a cooperative life-style—with insects. There can be little doubt that the crickets and other captive minstrels could recognize their owners, for they came to their call, as did Mrs. Alison Stillwell Cameron's. The owners would let their little charges out for a run on the floor before calling them back. The crickets seem to have made few attempts to escape. It is this kind of relationship between ourselves and the other creatures on this planet that we must strive to achieve.

4

INSECTS AS FOOD OR FOOD
PROVIDERS FOR MAN

If it weren't for our insect friends, we would find it very difficult to get anything at all to eat. The humble earthworm (see explanation of "insects" on page 11) breaks up the soil into small-enough pieces to let plants get rooted. Earthworms also soften and aerate the earth so the water and mineral food plants need to grow can reach them, and they do a magnificent job in breaking up decaying organic matter so that it can be utilized as fertilizer. All American earthworms, incidentally, are immigrants. They came to America in the dirt around potted plants brought by the first colonists. The native American earthworm population was killed off millennia ago during the ice age. (Of course there were plenty of other earth-burrowing insects around during the interval.)

Bees and other flying insects also ensure that we have vegetables and fruits—and grass on which the animals we eat can live. When Smyrna figs were first grown in California, the fruit growers discovered that the tree that bears the fig is female and has to be pollinated from the male tree, the wild caprifig. The Californian fruit growers accordingly imported these wild caprifigs and planted them by the female figs. Nothing happened, because there is a further link in the fertilization chain, a small wasplike insect called a chalcid fly. This insect lays its eggs inside a caprifig flower; but it will also settle on the flowers of the Smyrna fig. When it flies off, it leaves some of the pollen of the caprifig behind it, thus pollinating the Smyrna-fig flower.

Apart from producing most of our food supplies indirectly, insects themselves have entered into man's diet, though to a much smaller extent than might be supposed when we remember that there are more insects than any other living creatures. Various explanations for this neglect of insects as food have been given. It has been said, for instance, that a single insect provides very little food, and that man is too lazy to go around collecting all the insects he needs to make a good meal. Nobody who has watched French gourmets patiently extracting the shellfish from cockleshells would agree with this argument.

The fact is that we are prejudiced against insects as food. This prejudice probably goes back a long way, to the days when men first discovered that some of the butterflies were extremely poisonous. However, many insects can be eaten—and are eaten by some peoples. Generally speaking, insects are eaten either by people on the verge of starvation or by gourmets who like unusual dishes.

Explorers have often dined on insects, and Francis Galton, an Englishman who wrote a handbook for explorers in 1867, remarks: "Most kinds of creeping things are eatable and are used by the Chinese. Locusts and grasshoppers are not at all bad. To prepare them, pull off the legs and wings and roast them with a little grease in an iron dish, like coffee. Even the gnats that swarm on the Shire River are collected by the natives and pressed into cakes."

Prominent among those folk who were really starving before they began eating insect food were the children of Israel wandering in the desert. What they ate was not the bodies of insects themselves but a secretion produced from the insect's digestive juices combined with the sap of a plant, making a substance called manna. During the flight from Egypt, the Israelites had found nothing to eat. Most of them would have been glad to trade their newfound freedom for a square meal of the kind they used to get back in Egypt. Then manna appeared, spread out across the face of the desert like hoarfrost, and saved them from either dying of starvation or turning back. Think how different history would have been if the Israelites had stoned Moses to death (as they were on the point of doing) and gone back to the Pharaoh. What a different world we would be living in today.

What was manna? It is generally taken to be a substance of the kind I have described that can still be found in the Sinai Peninsula. During the months of June and July, gummy matter is exuded from the bark of the tamarisk tree or other bushes. These trees have been pierced by an insect of the scale-insect family, a creature called *Coccus maniparus*. The *Coccus* is a close relative of the same insect that produces cochineal. When it has pierced the bark of the tamarisk, a thick red syrup oozes out, sweet and rather like honey.

Manna used to be eaten by the wandering Bedouins of the desert and also by the monks of the monastery of Sancta Caterina on Mount Sinai. Two kinds of trees produce manna, *Tamarix mannifera* and *Tamarix gallica*. The supply of manna was very irregular. In some years there was none at all, whereas in others the Bedouins collected about seven hundred pounds. The honeylike "manna of Mount Sinai" used to be sold by the monks to the Russian pilgrams who were such frequent visitors to the Holy Land. In the cool hours of the morning, it solidifies and is formed into cakes, which fall from the branches. Powdered manna of this sort, blowing across the face of the desert, would certainly look like hoarfrost—broken up and powdery and mingled with the dust of the desert.

Manna is not confined to the Sinai desert. The learned rabbi David D'beth Hillel, who traveled from Jerusalem to India in 1832, noticed at Sulaymaniyah in Kurdistan (in what is now Iraq):

> There is to be found here, in summer, manna which comes every morning by the dew. That which falls on the rock is as white as snow, but it is very scarce to get, for it is taken for the governor and the nobles; that which falls on trees or grass is white and green, because it becomes united with the leaves and grass. This is found in great abundance. They sell it made up into balls. That which remains in the fields at the rising of the sun becomes as water. I have eaten it myself, it is sweet and of a pleasant taste and is used by the people as a medicine. The name of it in the native language and in Arabic is *mann shemma*, which means "heavenly manna."

The other insect food of the Israelites, the locust, has always been *the* insect delicacy most favored by gourmets. In the Bible, Leviticus, in discussing which creatures are lawful for the children of Israel, allows locusts to be eaten: "Even those of them ye may eat; the locust after his kind, and the bald locust after his kind . . ."

There are a number of references in the Bible to different kinds of insects, and it has been surmised that all of these refer to just one insect, the locust, in its different stages—from nymph through wingless larva to swarming insect. In living on locusts and wild honey, John the Baptist was probably following the example of the Prophet Amos, who lived on the same fare.

But besides being the fare of austere prophets, locusts have always been relished as a delicacy by kings. An alabaster relief from Kuysanjaq (Iraq) shows servants bringing locusts spitted on twigs for the supper of the Assyrian ruler King Assurbanipal. Another famous oriental monarch was a great locust eater, the Caliph Omar. Just as Moses permitted the locust "after its kind" to the Jews, so Muhammad allowed Muslims to eat locusts—rather surprisingly, in view of the fact that they are not allowed to eat shellfish. Locusts and fish are the only creatures that Muslims can eat unskinned, but they must have been killed by a Muslim.

"What do you think about the lawfulness of locusts as food?" someone asked Omar. "I only wish I had a basketful of them here," the caliph replied. "Wouldn't I scrunch them!" When there were no locusts to be had in Egypt, Omar sent messengers into Yemen, Cham, and Iraq, to try to collect some. When eventually one of these royal envoys brought back just a handful, Omar exclaimed: "Allah be praised!" The caliph's concern about locusts may have been not merely because they are good to eat but because there existed in the Arab world a superstitious belief that if locusts died out, men would go as well.

A large part of the Arab cuisine seems to have been centered around locusts. They were eaten fresh or preserved. They were grilled, boiled, or trimmed, and prepared in the *couscous*, which is such a delicious dish. Sun-dried or roasted between two layers of hot ashes, they were fed to camels, which relished them. Dried and salted, they formed an important element in commerce. They were exported on camelback all over the Muslim world, and the arrival of a large consignment of locusts in Baghdad would bring down the price of the meat in the market there.

Some travelers in the desert have described the locust as a wholesome and even a fattening food, which tastes a little like crab. Others have deprecated its nourishing qualities. Strabo, the Roman geographer, thought it was positively unwholesome. He describes a people who live on the borders of Arabia, the *acridophagi*, or locust eaters. These insect eaters procured their locusts by waiting till the equinoctial winds had begun to blow, assisting the locust in its migration. They would then light large fires and the smoke of the fire would blind and suffocate the locusts, which fell to the ground. The locust eaters ate the insects fresh or salted. "These *acridophagi*," says Strabo, "are, it is true, active and good runners, but their life never exceeds forty years." He attributes this short span of existence (which, after all, was not very short for the ancient world) to their peculiar diet.

Next to the locust there has been only one other great insect delicacy, the *palolo* in Samoa. The *palolo* is a marine worm (related to the annelids), which appears on the surface of the coral reefs of Samoa, Fiji, and Tonga on just one day of the year for a period of three hours. It is esteemed the greatest of all South Sea delicacies, and the islanders have always traveled long distances to get to the *palolo* reefs at the right time. Savage battles took place in the past when the first comers to a reef tried to prevent others from landing to catch *palolo*.

The sages of the island villages could reckon when the worm was due by noticing when the *aloalo* came into flower and by watching the constellations, particularly Orion. Just in case the old men of the village made a mistake, the principal "talking man," or orator, of each village in Samoa kept a dead reckoning. He hung a small basket on one of the rafters of his house. Into the basket he dropped each day markers—black pebbles, red and green feathers of the island parakeet, and leaves. The marking system was rather complicated, but when the basket held three leaves, five feathers, and four pebbles, the *palolo* was due the following morning. That night everybody went to bed early except for one wake-up man. At three o'clock in the morning, he roused the sleeping villagers with his call. They rushed down to the beach, carrying their paddles, and were soon over the reef, with the tide ebbing away beneath them. Jockeying for position, they tried to stay afloat in their canoes, with enough water beneath them to enable them to fish the *palolo* as soon as it came to the surface. As the tiny threadlike worms made their way up to the surface of the sea, the Samoans scooped them up with small nets on handles and ate them raw.

So delicious was the *palolo* that its arrival marked the beginning of the Samoan year. An adventurous American lady called Llewella Pierce Churchill, who lived in Samoa at the turn of the century, describes the taste of the *palolo* as being like that of the sweet scallop. Wrote M. L. Budgen:

> While John the Baptist was subsisting in the desert of Judaea on the simple and ordinary fare of "locusts and wild honey," imperial, luxurious Rome was regaling in her banquet halls upon veritable insects, luscious caterpillar grubs fattened on flour, as we fatten oysters upon meal. This was the *cossus* of Pliny, and was supposedly identical with the unsightly, wood-devouring larva of the Great Goat Moth, a lurid red and yellowish caterpillar, bulky, black headed, and black clawed, a darkling dweller in the trunk of oak or willow.

The larva of the goat moth—all stages of its life are shown here—a black-headed, red-and-yellow caterpillar, was considered gourmet fare in ancient Rome.

Caterpillars and other larvae have probably been the most favored insect foods, next to those I have already mentioned. Even nowadays, central African miners working in South Africa have dried caterpillars sent to them as a great delicacy. Larvae have been so popular with some people that they have given them their national name, like the Australian aborigine tribe called the Witchetty Grub people, after their favorite food.

All sorts of larvae have been eaten, including the palm-weevil grubs relished by many Africans. When some of these Africans arrived as slaves in Mauritius and the West Indies, they taught their masters the taste for this grub. In the same period upper-class Englishmen enjoyed the mites found in a Stilton cheese and encouraged them by pouring in a spoonful of port every time they took a slice out of the cheese.

Section through a termites' nest, or "castle." In a well-protected spot in its center is the underground royal chamber (*center right*), where the king and queen spend their entire lives once they mate. In time, the queen (*foreground*) becomes an immovable but highly efficient egg factory, with a grotesquely huge abdomen a thousands times the size of the workers, who swarm over her to feed and groom her constantly. Termites are considered a delicacy among Africans. A food gatherer (*left*) tries to break into a nest with a stick.

Some American Indians relied on insects very much for their food. They ate brine flies, waterbug eggs, and stinkbugs ground up with chili or pimento pepper, as a relish on tortillas. The Digger Indians of California were particularly fond of grasshoppers. They drove them into pits, dipped them in saltwater, and baked them in ovens. "Having from curiosity tasted of the roast," writes a reporter on the *Empire County Argus*, "really if one could divest himself of the idea of eating an insect as we do an oyster or shrimp, without other preparation than simple roasting, they would not be considered very bad eating, even by more refined epicures than the Digger Indians."

Ants have always been an important food-producing insect. In Africa, the Hottentots used to eat termites boiled or raw, and other Africans roasted them in iron pots and ate them by the handfuls, as we do jelly beans. According to one traveler called Smeathman, they resembled sugared cream or sweet almond paste. In Japan, chocolate-coated ants are eaten as a delicacy, and even in Europe, where the prejudice against insects as part of the diet has always been strongest, the Swedes used to distill ants with rye, and the distillate was added to brandy to flavor it.

Ants are eaten in both East and West. The Japanese dip them in chocolate and consider them party fare. The Swedes distill them with rye.

The most important insect contribution to food and drink has always been the honey produced by the bees. John the Baptist was able to subsist on a diet of wild locusts and honey alone. Man has apparently always had a sweet tooth, but until sugarcane was planted in the West Indies in the sixteenth century, it was almost impossible for him to satisfy his taste for sweets, except by eating something flavored with honey, such as the honey cakes that the ancient Egyptians loved so much.

The quantity and importance of food provided for human beings by the bees was incalculable. In Russia a Ukrainian peasant might own five hundred beehives; one Spanish priest had a thousand. But not all honey was used for food. A very large amount was turned into a fermented drink called mead. Mead was the principal tipple of the upper classes in ancient Europe. Wine was incredibly expensive, the cultivation of the wine then being confined to Greece, Italy, Egypt, and North Africa. In ancient Gaul an *amphora*, or jar, of Greek wine could be readily exchanged for a male slave. So the ancient Gauls, British, Scandinavians, and Germans were forced to drink beer or mead. Mead was made by boiling honeycombs, from which the honey had been extracted, in a caldron of water. This decoction of honey and water (as well as other bee foods, such as pollen, royal

Man's favorite food-producing species, far and away, is the honeybee. Here a nineteenth-century keeper collects a swarm in a skep, a straw-woven hive.

jelly, and so forth) would be tipped into great casks (large enough to contain a human being, as will be seen) and left to ferment. The result was a frothy, golden, fiery liquor with lots of alcohol in it. Even nowadays Scandinavians regard mead as so potent that it is banned in Iceland.

A love of mead was the one characteristic that the old Celts and the old English inhabitants of Britain had in common. Divided in all else, they united to praise the virtues of this divine vintage of the bees. The Welsh bard Taliesin sang:

May Maelgwyn of Mona be inspired with mead,
May he cheer us with drink.
From the mead horn's foaming, pure and shining liquor,
Which the bees provide, but do not enjoy themselves.
I praise mead! Its praise is everywhere.
It is precious to all the creatures which all the earth man maintains.
God made it for Man, that he might be happy.

Unfortunately the very last thing that the ancient Celts or old English did with their mead was to allow it to make them happy. Instead, a bout of drinking mead was often a preliminary to a brawl—and sometimes a real battle in which lives were lost. Fuddled with their mead, the English fell easy victims to a sudden attack by their enemies. They would be dragged from their mead benches and slaughtered.

Mead was drunk in a special building called a mead hall. In the oldest of all English poems, *Beowulf*, the author, speaking of a king called Hrothgar, says:

> *It ran through his mind*
> *That he a hall house*
> *Would command,*
> *A great mead house*
> *Men to make*
> *Which the sorts of men*
> *Should ever hear of.*

Inside the mead hall were long benches, on which the mead drinkers sat. The drinkers on these mead benches were socially superior to ordinary men, who drank beer. Mead was drunk not from a goblet or glass but from a *horn*, usually the horn of an ox or the ivory tusk of a walrus. They were often beautifully carved, and some had special stands on which they could be set when not in use. Quite a number of these horns have survived, notably one belonging to a man called Ulph, which has been preserved in York Minster, the cathedral in York, England. Though a sober man himself, the American poet Longfellow was so inspired by Anglo-Saxon drinking horns that he wrote a poem about one of them, which had belonged to Witlaf, king of Mercia. These beautiful drinking utensils had a fault that many other drinking cups of the ancient world shared. They held too much. What was more, there was no way of putting down a horn of mead if it contained any liquor—you had to drink it straight off. Then it would be immediately refilled by your host. The horns were so heavy that on occasion they were used as clubs, and drinkers knocked one another down.

The most important date in English history is A.D. 1066. That was the year when William the Conqueror invaded England and managed to defeat the last Anglo-Saxon king, Harold, in the battle of Hastings. William might never have won but for the fact that Harold's army was exhausted. They had just marched south, at great speed, after defeating Harold's treacherous brother Tostig, who had invaded the north of England together with a Norwegian king, also called Harold. What had put the two brothers on different sides? They had had a quarrel, during which Tostig had retired to one of Harold's country houses and there slaughtered his household servants. This had been bad enough, but then Tostig had added insult to injury. He had put the dead bodies of Harold's retainers in the mead casks in the cellar and sent Harold an insulting message, telling him that he would find plenty of preserved meat when he came there. Not merely had Tostig ruined part of Harold's supplies of drink for the winter, he had probably ruined his pleasure in mead for life. The civil war between the brothers made it possible for Duke William and his Normans to slip across the Channel while it was unguarded. Thus, were it not for a quarrel over some barrels of mead, you might still be speaking Anglo-Saxon, the old form of our language, with all its declensions and conjunctions intact. Instead you speak English, a much simpler form, which evolved through the contacts of the Norman rulers of England and their Anglo-Saxon subjects.

After centuries of being forgotten about, mead is being made in England again. It is also the national drink of Ethiopia, under the name of *tej*. "No one can call himself a man," my Abyssinian guide told me, "unless he can drink *tej* and eat raw meat."

5

INSECTS IN MEDICINE

Insects, said the Victorian entomologist Budgen, were once highly regarded as a medicine—quite as much so as herbs.

To take a wood-louse or Millepied, perhaps alive, and conveniently self-rolled, for the occasion, was as common as to take a vegetable pill. Five Gnats were administered with as much confidence as three grains of calomel. In an alarming fit of cholic, no visitor with a dram of peppermint could have been more cordially welcomed or swallowed than a Lady-bird. Fly water was eye water, and even that water-shunning monster, Hydrophobia, was urged to lap *aqua pura* by the administration of a dry Cockchafer.

By 1849 the belief that insects had medicinal properties had been largely abandoned. Nonetheless the interest felt in insects as a source of medicine is very important historically, as it is one of the main reasons why humans first became interested in insects. The first mention of the cicada, for example, seems to have been made in a Chinese pharmacopoeia written about three thousand years ago. Insects were more usually employed in folk medicine than by regular doctors. People had found out, empirically, that some insect products could really effect cures. A good example is the spider's web. In England carpenters and anyone else working with sharp tools who cut themselves used to clap a spider's web onto the cut immediately. This stopped the bleeding, because the substance of the spider's silk contains a coagulent.

The most important insect in medical history is the leech. The leech is a kind of earthworm that attaches itself as a parasite to animals or humans. While in this parasitic state, it sucks blood from its host, gorging itself from his veins. It was precisely the blood-sucking qualities of the leech that doctors were interested in. They still are, because the leech is even now featured in the medical pharmacopoeia of England. However, it is still used for only one condition—a black eye. Nothing can relieve the congestion of a black eye faster than a leech, applied to the bruise until it has gorged itself.

When not drinking a patient's blood, leeches live in water in glass jars. When I was a small boy, I used to keep two of them as pets, having begged them of a pharmacist who was my next-door neighbor. I thus had a good opportunity to study the special kind of leech used in medicine in England, the *Hirudo medicinalis*. The leech looks like an earthworm with one end thicker than the other. Its mouth opens in three slits, edged with about eighty teeth, with which it can open the skin of its prey and suck its blood.

A few centuries ago leeches were prescribed, not just for black eyes, but for almost every conceivable ailment. Doctors believed that bloodletting—getting rid of "bad" blood—was a sovereign remedy for all ills. They believed in bloodletting so much that they themselves were often called leeches. The *Hirudo medicinalis* had become established in England, but there were never enough of them for English needs. Some men became professional leech gatherers and went about collecting them from ponds. The famous English poet William Wordsworth records his conversation with such a man in a poem written in 1802.

> *My question eagerly did I renew,*
> *"How is it that you live, and what is it you do?"*
> *He with a smile did then his words repeat:*
> *And said that gathering leeches, far and wide,*
> *He travelled: stirring thus about his feet*
> *The waters of the pool where they abide.*
> *"Once I could meet with them on every side;*
> *But they have dwindled long by slow decay;*
> *Yet still I persevere and find them where I may."*

There were plenty of these leech gatherers around, all looking for an ever-diminishing number of British leeches. In Ireland the local people used to gather leeches in Lough Mask and other inland lakes by sitting on the edge of the pool and dangling their legs in the water till the leeches had fastened onto them. Leech ponds were set up in England, such as one at Heacham in Norfolk. They were also raised in the United States. But the best leeches came from the continent of Europe. In France old horses were driven into the marshes and leech ponds to provide food for the leeches. A group of animal-loving Frenchmen managed to get this practice stopped, and instead the leeches were kept in ponds and fed on refuse from slaughterhouses. In 1832 alone, 57 million leeches were imported into France. They were transported from place to place in casks of clay and water, stone jars, or baskets full of moss, carried in special litters.

There can be no doubt that the leeches drew blood from patients as efficiently as doctors did by opening a vein—a practice called venesection. Some patients—often women—who did not like the doctor's lancet had leeches instead. Many prominent women in history died as a result of overbleeding, such as George IV's daughter, Princess Charlotte, who died in childbirth. The death of this unfortunate princess left the throne open for Queen Victoria.

Leeches became one of the largest insect industries in history. By the late nineteenth century, 16 million leeches were employed yearly in England alone. In France, consignments of between 60,000 and 80,000 used to leave Strasbourg for Paris. In 1806 a thousand leeches in France fetched twelve to fifteen francs, but in 1821 the price had risen to as high as 283 francs.

Often people who acquired leeches for medical purposes then kept them as pets, because they felt that they could foretell the weather. Leeches were kept in "leech barometers." These were simply glass jars, closed at the top with some porous cloth and kept half filled with water. When a storm was imminent, said leech owners, the leeches would climb to the top of the jar and move about with agility. Other owners kept their leeches out of gratitude. Lord Erskine, a famous English Lord Chancellor, was a case in point; a friend described Erskine's explanation:

> He told us how that he had got two favorite leeches. He had been
> blooded by them last autumn, when he had been taken dangerously ill

at Portsmouth; they had saved his life, and he had brought them with him to town, had ever since kept them in a glass, and himself given them fresh water, and had formed a friendship with them. He said he was sure they both knew him and were grateful to him. He had given them different names, "Home" and "Cline" (the names of two celebrated surgeons), their dispositions being quite different.

All the other activities of insects in medicine yield in importance to the parts played in providing us with the first modern anesthetic, chloroform. This anesthetic was, in fact, a mutated form of formic acid, the liquid produced by ants. (The word "formic" is derived from Formicidae, the ant family.) Two seventeenth-century English naturalists, John Ray and Dr. Martin Lister, had done much research on this acid, which they obtained by crushing ants in a mortar with a pestle. Ray and Lister could not be certain about the purpose of this acid. We know now that it was intended as a sort of signpost for the ants, helping them to find their way around during their travels. If you doubt this, get a bottle of formic acid and a brush, go near an ant's nest, and write your name in large letters on the ground. The ants, attracted by the smell, will come out of the nest and crawl up and down all the letters until your name is spelled out in live ants.

The experiments of Ray and Lister attracted the attention of a French chemist called Dumas. Dumas broke down the acid and found that it contained a base, a compound radical, which he named formyl. This base, with three atoms of oxygen, forms formic acid. Dumas discovered that he could replace these three atoms of oxygen with three atoms of chlorine. He thus obtained trichloride of formyl. Once ether had been employed as an anesthetic, Dr. Simpson of Edinburgh started looking around for an improved substitute. He tried trichloride of formyl, experimenting on himself. The results were sensational. Simpson, and later his assistant, was completely knocked out by the substance and did not come to until hours later. Trichloride of formyl became famous under a new name, "chloroform," and the second half of this name recalls our useful friend, the ant.

The oldest of all insect medicines is probably cantharides. Even now it has not completely disappeared from the British pharmacopoeia and is used in hair restorers and other lotions. For over two thousand years—and perhaps much longer—*cantharidin*, which is the active principle of cantharides, has been used both externally and internally. Aretaeus, a Roman physician of the first century, prescribed it for blisters. Hippocrates, the Greek "Father of Medicine," who lived around 400 B.C., administered it as a cure for dropsy, apoplexy, and jaundice. Besides raising blisters, cantharidin is supposed to be an aphrodisiac, but again and again through history, this drug has caused fatalities. What was administered as a love potion sometimes turned out to be a murderous, or suicidal, dose. Cantharidin is obtained from the blister beetles (family Meloidae), especially from *Lytta vesicatoria*, the Spanish fly.

This creature is not a fly at all but a handsome green beetle about four fifths of an inch long. It is found on ash, lilac, privet, elder, and other trees, and is particularly common in Spain, hence the name. Since the Spanish flies gather together in large numbers, it is easier to collect them than other members of the same family. The cantharides give off a strong smell of mice, which helps the gatherers detect them.

Cantharides, or Spanish fly (actually a beetle), is collected by shaking the branches of the tree where it settled. A touch of it will blister, so it must be handled with care.

The beetles are usually found settled on an ash. Very early in the morning a muslin cloth is spread out around the tree. Then one or more of the gatherers climb up into the branches. The branches are shaken, and the beetles fall down onto the cloth. They are numbed with cold and do not try to escape. When enough beetles have been gathered, the collectors draw the cloth together, working very gingerly, because the touch of one of the beetles will blister a man's hand. Once the whole cloth has been bundled together, it is plunged into a tubful of hot water and vinegar, which kills the beetles. Their bodies are then dried in lofts, or on sheets of paper spread out under an airy shade. When dry, the insects, which still must be handled with gloves, are packed in airtight boxes or in glass or earthenware vessels.

The action of raising a blister on the skin was felt to be essential for "purging away humours." Blisters were also applied in veterinary surgery to horses. It is perhaps difficult to see how a drug that can produce blisters, cause vomiting of blood, renal failure, and extreme depression has had such a long life in medicine. However, during the Civil War, cantharidin was considered such an important drug that, when supplies from outside were cut off by the Federal blockade, Confederate doctors employed as a substitute the Colorado potato beetle, *Leptinotarsa decemlineata*. The latter is the best-known American insect in Britain, for its photograph can be seen outside any police station, with a reward offered for its discovery in any potato field. It is a very much dreaded pest in that country today.

The story of insect remedies in medicine by no means ends there. Insect toxins have been used experimentally for the cure of rheumatism, an ailment from which beekeepers rarely suffer. Even the noxious mosquito has turned out to have a useful side. Malaria mosquitoes are used to infect patients suffering from a dreadful condition called paresis, a final stage of syphilis. The malarial organisms deal with those of syphilis by eating them up. All that then needs to be done is to cure the malaria.

Generally speaking, insects have always entered more into folk medicine than into the prescriptions of orthodox physicians. They are widely represented in the traditional medicine of China, where *Mylabris chicorii* is used for producing blisters, and in that of Africa. This insect, known as the Telini "fly," extends in range from Italy to China. It is very rich in cantharidin, yielding twice as much as

ordinary cantharides. During my stay at Blantyre, in Malawi,
Central Africa, I was always enchanted, whenever I visited the
market, by the varieties of organic substances the pharmacopoeist
there had on display. There were rhizomes, dried snakes, seeds,
herbs and roots, and the chrysalises and other parts of insects,

Bushmen killing ostriches with poisoned arrows. One of the hunters tries to
approach the flock hidden under the plumage of a dead bird with the head
still attached.

together with whole dried millipedes. Western observers, even
trained physicians like Dr. David Livingstone, have often been
impressed with the African's knowledge of drugs. In Livingstone's
day, a very much dreaded branch of African pharmacopoeia was the
knowledge of poisons, for which the Bushmen were famous. The
Bushmen (who are now virtually extinct, apart from survivals in the
Kalahari and in Namibia) were able to keep their enemies at a distance
for centuries by firing at them arrows tipped with a poison produced
by a grub called *n'gwa* or *k'aa*. The poison grub, which lives on the
branches of a tree called *Maruru papeerie*, is a pale flesh color about
three quarters of an inch long. All the internal organs of the grub
contain the most deadly venom, and to poison his arrows the
Bushman first examined his fingers to make sure that the skin was not

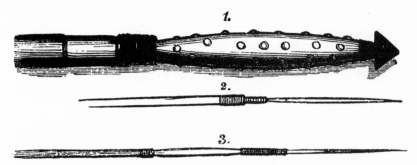

1. Barbed arrowhead, full size. 2. Unbarbed arrowhead in shaft. 3. Unbarbed arrowhead with poisoned point outward.

broken, then took a grub and squeezed out all its juices onto the point of his arrow. The effects of this poison were terrible to experience, or even to see. In Livingstone's words: "The sufferer cuts himself, calls for his mother's breast, as if he were returned in his idea to childhood again, or flies from human habitations as a raging maniac."

Not only did the Bushmen shoot poisoned arrows, they poisoned the waterholes behind them as they retreated. Moffat, the father-in-law of Livingstone, was poisoned by drinking at one of these ponds, but luckily survived.

6

INSECTS IN ART

---

*A*t first sight it might appear as though there were nothing in nature so inartistic as an insect. Many people still regard insects as nasty, creepy, unpleasant creatures. One expert on porcelain claims that you can never sell porcelain decorated with insects, for people will not want to use it at table. In spite of this prejudice, insects have had a very stimulating effect on art. Take cosmetics, for example.

Many primitive people employ body paint—like the ancient Britons who were reputed to dye themselves blue with woad. The Maoris of New Zealand, like the American Indians, painted themselves to go to war. But an early observer of the Maoris noted that paint was "used in peace as well as war, being regarded as a good preservative against the bites and stings of insects, especially the sandflies and mosquitoes."

The American Indian used peacetime paint, too, and for the same reason. The Earl of Carlisle, who had come to America to try to arrange peace between the rebels and George III, wrote, with pardonable exaggeration, "The mosquitoes here are the size of sparrows. I have given up wearing knee breeches and silk stockings and now wear trousers to avoid their bites."

The use of face paint throughout history has been almost universal. In ancient Egypt, a slave was appointed to anoint the guests when they arrived at a banquet. The slave put a pat of scented ointment, which he took from an alabaster jar, on the head of each guest as he took his seat. The pat of ointment slowly melted away, allowing the perfume to run down the guest's face. This was what the Bible means by "the oil of joy." Other African people besides the Egyptians anointed themselves with grease—no Bantu would think of setting out without a pat of butter on top of his head, which protects the skin against the rays of the sun. But the Egyptian ointment was intended to protect the guests against the horrible insect pests that are rampant in Egypt, particularly those very determined African flies that try to get right into a person's ear.

Thus the art of face and body painting, originally employed as a defense against insect bites, became in time a way of enhancing beauty: the science of cosmetics was born.

The extraordinary strength displayed by insects—their capacity to carry or pull much more than their own weight, which was first investigated by the French entomologist Plateau, gave us the delights of the flea circus. This unique form of entertainment originated in the Far East, where insects are so much a part of life. William Elliot Griffis, an American who taught science in a Japanese college between 1870 and 1874, describes the amusements for children that could be seen in Yanagi Cho (Willow Street) in Tokyo at that time. "The bug man," he says, "harnesses paper carts to the backs of beetles with wax, and a half-dozen in this gear will drag a load of rice up an inclined plane."

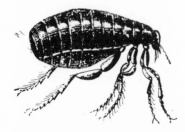

The lowly flea, bane of dogs and man, is nonetheless a circus entertainer of remarkable intelligence and skill and learns his trade fast.

Fleas were not being used in this Japanese circus, but when the insect circus emerged in Europe, it was the strength, agility, and intelligence of that tireless insect, the flea, that made it the star. Feats that men find extremely difficult seem easy to a flea. Were a flea man-size, it would have no difficulty in jumping over Grant's Tomb. As early as the sixteenth century, showmen began to exploit this unpaid actor, *Pulex irritans*. They harnessed it with a minuscule collar, sometimes of gold, attached it to some kind of theatrical prop, and set it to work doing various tricks, usually in a box whose lid was a magnifying glass. At the end of the day the flea got its revenge by being allowed to bite its keeper. "I lives off them, and they lives off me," an English flea-circus proprietor used to explain to the crowd.

Circus fleas—poor insect slaves—could be bought at Venice and Augsburg, "and at a small price too." They were sold, already harnessed, with steel or silver collars. They had to be kept in a box among wool and fed by their owner once a day. Once lassoed and coralled, the flea showed remarkable intelligence and adaptability, and understood what was required of it much quicker than the average horse does.

Thomas Moffat, a contemporary of Shakespeare who wrote the first book on entomology in English, was very much impressed with an English flea circus that he attended:

> A certain Mark, an Englishman skilled in all subtle craftsmanship, attached a chain of gold, a finger's span in length, together with a lock and key to a flea, and with so much ingenuity and subtility that it could be easily pulled behind the flea as it went along. Yet the flea, the chain, the lock, and the key, did not exceed a grain in weight. We have also heard, from trustworthy people, that a flea harnessed in this way with a chain, without effort pulled a gold coach which was perfect in all its details, something that greatly commends the industry of the craftsman, and the strength of the flea.

The craftsman whom Moffat mentions was called Mark Scalliot. A later English craftsman, a Mr. Boverick of the Strand in London, was a watchmaker. Sometime before 1847, he exhibited "a little ivory chaise with four wheels, and all its proper apparatus, and the figure of a man sitting on the box, all of which were drawn by a single flea." Mr. Boverick also constructed a minute landau (another kind of carriage) which opened and shut by springs, with the figures of six horses harnessed to it and a coachman on the box, a dog between his legs, four persons inside, two footmen behind it, and a postilion riding on one of the fore horses, which were all easily dragged along by a single flea.

At a nineteenth-century fair in Charlton, in Kent, England, a man exhibited three fleas harnessed to a carriage in the form of an omnibus. The omnibus was at least fifty times the bulk of the fleas, but they pulled it easily. The Abbé Latreille, whose life was saved by a beetle (see p. 125), saw another flea, which dragged a silver cannon twenty-four times its weight, mounted on wheels, and did not show any sign of alarm when the cannon was loaded and fired.

In 1825 a Baron Walckenaer was one of the many Parisians who joined the line to pay his sixty centimes and watch "The Learned Fleas" exhibited in a room on the Place de la Bourse in Paris. The spectators sat down in front of a curtain, which had openings in it. Through these openings they could see the various "rings" in which different acts of the flea circus were being performed.

> Two fleas, [says the baron] were harnessed to, and drew, a golden carriage with four wheels and a postilion. A third flea was seated on the coach box and held a splinter of wood for a whip. Two other fleas drew a cannon on its carriage. This little trinket was admirably finished, not a screw or nut was wanting. These and other wonders were performed on polished glass. The flea horses were fastened by a gold chain attached to the thighs of their hind legs, which, I was told, was never taken off. They had lived thus for two years and a half, not one having died during the period. To be fed, they were placed on a man's arm, which they sucked. When they were unwilling to draw the cannon or the carriage, the man took a burning coal, and on being moved about near them, they were at once aroused, and recommenced the performance.

Even nowadays, the repertoire of a flea circus, if one could be found, would be little different from what has already been described. Flea shows tended to appeal to a conservative audience. The public knew what to expect and liked what it saw. Moreover, the flea circus, if brought to perfection first in England, had soon become international, drawing on artists from as far away as Italy and Germany.

The only other live insect artists were the *cucujos*, or firebugs. One of these Coleoptera, whose Latin name is *Pyrophorus noctilucus* ("the fire-bearing night light") once made his presence felt far from his native America—in Paris, in fact. At twilight one evening the firebug, which had apparently arrived in Europe by accident in a consignment of timber from Mexico or Guyana, suddenly appeared in the Faubourg St. Antoine. As the insect "star" flew glowing over their heads, the inhabitants of the quarter shrieked with terror and ran for their lives.

The *cucujos* had long been used by Indians and Spanish Mexicans as a light to the feet and an adornment for the person. In order to catch the firebugs, the Indians in the West Indies went out of their huts just as the speedy tropical twilight began. In their hands they carried a blazing torch, which they swung around and around. They climbed

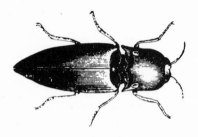

The *cucujo,* or lantern beetle, gave such a brilliant light that the natives of Mexico and Central America used them as flashlights and living jewels.

up a small hillock so that as many insects as possible could see the flame, and cried, *"Cucuji, cucuji."* Attracted by the light—and as the Indians felt, by the call as well—firebugs hastened to the spot. The Indians used them as living torches, fastening them to their sandals or to their great toes, so that they could hunt rodents in the forest at night. When morning came, they were always very careful to untie the *cucujo* and replaced it in the wild, on a suitable bush.

The light from a number of *cucujos* was so brilliant that to the feverish eyes of the sentries of the conquistadores Nárvaez and Cortés it seemed like an army of musketeers advancing with their slow matches already lighted, ready to fire their guns.

Cucujos had long been worn by the ladies of Mexico as living jewels to adorn them at the ball. The Indians collected the insects in the forest and sold them to the Mexican belles, who kept them in little cages of very fine wire, feeding them on fragments of sugarcane and bathing them twice a day to refresh them. Entomologist Louis Figuier describes the adorning process thus:

When the Mexican ladies wish to adorn themselves with these living diamonds, they place them in little bags of light tulle, which they arrange with taste on their skirts. There is another way of mounting the Cucujos. They pass a pin under the thorax, without hurting them, and stick this pin in their hair. The refinement of elegance consists in combining with the Cucujos, humming birds and real diamonds, which produce a dazzling headdress. Sometimes, imprisoning these animated flames in gauze, the graceful Mexican women twist them into ardent necklaces, or else roll them round their waists, like a fiery girdle. They go to the ball under a diadem of living topazes, of animated emeralds, and this diadem blazes or pales according as the insect is fresh or fatigued. When they return home, after the *soiree*, they make them take a bath, which refreshes them, and put them back again into the cage, which sheds a soft light in the chamber during the whole night.

Scarcely less showy and brilliant than living jewels were ornaments made from the wing cases and other parts of beetles. The gold-band beetle (*Buprestis vittata*) was used in India for adornment, and its gold-striped wing cases were also used to embroider slippers and dresses. In the East Indies, China, and even the south of France, *Buprestis* ornaments also appeared in women's hair arrangements and were used to make earrings. In Chile and Brazil, ladies made necklaces of the gold beetles, or Chrysomelaidae, and also from the brilliant diamond beetles. An allied species of insect was used to make the finest Parisian artificial flowers. Many elytra (wing covers) of beetles appeared on fans, both in India and in Europe, where the old-fashioned peacock fan was also ornamented in this way. The French naturalist Réaumur remarked that "diamonds have perhaps beauties no more real than those of a butterfly's wings."

In the first half of the twentieth century, the *Morpho* butterflies of South America were farmed so that their glistening wings could be employed for various ornaments. In China, large and beautiful butterflies were caught on Lo-fu-shan Mountain, in Kwantung Province, and sent to the court of the emperor, where they were made up into ornaments for the palace. In Siberia, butterflies were caught by political exiles in Czarist days and sold to dealers. Convicts in the French penal settlements in the Pacific and South America were also allowed to collect butterflies.

Sometimes, instead of being made into jewelry, parts of butterflies are formed into complete pictures by building up pieces of wings into a mosaic. West Africa is particularly noted for these butterfly mosaics. Other species of insect have been used extensively in art. Vinh Thai, a member of a Free French Mission, described how, while on duty as a diplomat in Chungking in 1943, he saw on the stall of a toy seller an entire landscape inside a glass snuff bottle. The

Not even the cockroach has escaped being useful to man. This hardy survivor from a remote past has been known to supply the raw material for artists. Cockroaches are also valued as food in many parts of the world.

rocks were made from pieces of eggshell and the trees were constructed out of cockroaches' legs.

A final and rather macabre use for insects occurs in English folk art of the nineteenth century. A memorial inscription would be written out on paper, commemorating a departed friend, perhaps with a few elegiac verses and always with the name of the friend and the date of his or her death. This calligraphic exercise would then be mounted on a board and surrounded with lines of decorative beetles or other insects, pinned in rows. Today these insects have now virtually all fallen to bits, a much stronger reminder of the transience of everything earthly now than when they were first mounted.

It is a small step from jewels made from insects to jewels made to look like them. Insects provided the inspiration for the most important of all antique gems—the scarab. Just to mention the word "scarab" conjures up the departed glories of ancient Egypt. Yet many non-Egyptian people copied the scarab, and it has been found in Judea, Phoenicia, Cyprus, Etruria, and many other parts of the ancient world. All this artistic inspiration came from just one small beetle, the *Scarabeus sacer*, one of the family of insects called Scarabaeidae.

These beetles have a blackish-green metallic luster that makes them look as though they had been dipped in enamel. The peculiarly shaped hind legs of the scarab, placed far away from one another, give it an odd appearance when it crawls around. Yet these legs are very well adapted to the scarab's main task, rolling up pellets of animal dung into balls and trundling them off into holes where the beetle lays its eggs; when the eggs hatch, the pellets serve as food for the larvae. Obviously the scarab is doing a useful job as a scavenger, because nothing breeds flies like animal dung left lying around in hot countries, where there are enough flies to begin with.

The ancient Egyptians, who venerated the scarab, did not do so because it was useful to man, but because they felt it had godlike attributes. The religion of the Egyptians included veneration of many animals, birds, and reptiles, but the scarab was the only insect they worshiped. They observed the strange habit of the scarab's rolling up its ball, but misinterpreted them. Not knowing there were male and female scarabs, the Egyptians believed that the scarab beetle created life from the earth by simply rolling up its dung pellet. These dung pellets were taken to be a symbol of the round earth and also of the sun, which, according to Egyptian belief, rolled across the heavens. Thus this beetle came to symbolize resurrection and rebirth, and so the scarab became synonymous with anything that was unique or self-creating. All these ideas were personified in the god Chepera, who had a beetle in place of his head. One of his tasks was to roll the sun across the horizon.

Scarab beetles roll up pellets of dung in which to lay their eggs. This curious habit made the ancient Egyptians believe that the sun was revolved by an enormous scarab and hence that the beetle was lucky.

The scarab also became the emblem of immortality. One reason for this was that the scarabaeus seemed to be indestructible. It appeared on the fields and along the edges of the desert, especially once the crops had been cleared. Shortly afterward the Nile rose and flooded the whole country. Yet when the Nile withdrew its floods from the inhabited land of Egypt and everyone came down from the mounds on which the villages were built, or from the stony edge of the desert, there were the scarabs again, apparently none the worse.

The scarab became one of the most popular forms of jewelry. Even nowadays, the soil of Egypt is literally sown with scarabs in some places. When a few tons of Egyptian earth were sent to England for examination by soil specialists, the specialists said rather snootily that Egyptian earth was no different from that of any other country. The earth was handed over to the parks department, who used it as top soil in community housing projects. Almost immediately people started pouring into the British Museum, showing scarabs they had found in their backyards.

Most scarabs (the jewel) are fairly small, from about half an inch to an inch long. They were cut so that the back of the stone looked like the back of the scarab beetle, and the underside was used for inscriptions. Scarabs were made from all the precious stones known to the Egyptians: jasper, amethyst, cornelian, ruby, lapis lazuli, and others. They were also made in decorative stones, such as green basalt, granite, steatite, and others, or in composition such as pottery or faience—a paste that hardens into a beautiful colored glaze, usually blue in ancient Egypt. Scarabs were also made in gold, mostly for pharaohs. Many scarabs had a hole bored through them, by means of which they could be attached to a ring or suspended from the owner's neck with a thread or chain.

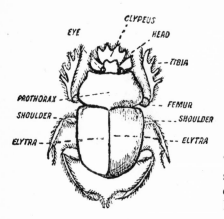

The anatomy of the scrab beetle, shown here, formed the basis for the decorative scarabs so much prized by Egyptians.

The quality of a scarab depends much more on how old it is and how well it is cut than on the material of which it is made. The scarab in the Egyptian gallery in the British Museum, for example, is priceless, even though it is only made from basalt, because it is the biggest scarab in the world, and it would take a fork-lift truck to move it. The best scarabs give a true representation of the beetle, carefully carved and showing the head, prothorax, and elytra, as well as the fold of the wings, which is sometimes outlined in gold wire. The legs and tibia of the scarab are usually rather sketchily carved or left out altogether. Because scarabs are made in many different colors and sizes, a collection of them in a case looks very decorative. Some of the best scarab collections are found in the United States.

The collector and the Egyptologist are not so much interested in the appearance of scarabs as in what they have to tell us. Although a few scarabs are merely engraved with a few decorative devices, most of them are inscribed in hieroglyphics, the written language of ancient Egypt. The inscriptions on the undersides of scarabs usually give the names of gods, priests, pharaohs, officials, and private persons. Because scarabs are made from a hard stone or pottery, they have survived where everything has been lost. Scarabs can teach us a lot about ancient Egyptian history. In fact, several pharaohs are known to us only because one of their scarabs has survived.

Scarabs were made for quite a number of purposes. Some were created to commemorate a special historical event. One historical scarab was engraved on the order of Amenhotep III, a pharaoh who ascended the throne of Egypt in 1417 B.C. Amenhotep was one of the greatest of all Egyptian rulers. Like many men of action, he evidently realized the place that sport should play even in the busiest man's life, because that is what this scarab is all about. The inscription describes how, in the second year of his reign, a messenger came to tell this pharaoh that something unusual had occurred. Wild cattle, something very rare outside the deep desert, had been sighted at Shetep. Boarding his river barge called *Shining-in-Truth*, Amenhotep went downriver. The whole Egyptian army had been marshaled as beaters, and they drove the wild cattle, numbering 190, into an enclosure marked off by nets and dikes. Amenhotep himself then proceeded to kill forty of them with his bow.

Historical scarabs such as this one are very rare; much more common are funeral scarabs. These were usually engraved with a passage from the *Book of the Dead* and laid on the breast of the mummy in its tomb. Then there were personal scarabs, which were the most common. They were often worn as rings, for personal adornment, or for good luck. Some of them bore the name of a god or of a famous pharaoh, long dead. If the scarab bore the name of the owner, he might us it to stamp the seals of papyrus documents. After a man had witnessed the writings, they were rolled up, tied with a thread, and a blob of soft wet clay was placed on the knot. Then he would press the clay seal with his scarab. A scarab might also be used to seal anything that someone wanted put away or kept intact. This could be a jewelry box, tied up with a cord, the door of a treasury or of a tomb, or even a jar of special wine. Scarabs were often mounted on rings in such a way that they could swivel around. The back of the beetle would be displayed on the finger, but if the ring was wanted as a seal, instead of as an ornament, the scarab could be turned around so that the inscription side was uppermost.

One of the most important uses of these signet scarabs was as emblems of authority. We read in the Old Testament how "Pharaoh took off his ring from his hand and put it upon Joseph's hand . . . and made him ruler over all the land of Egypt." A painting from an Egyptian tomb shows this kind of investiture taking place. It shows the chancellor of Pharaoh Tutankhamen, who lived around 1350 B.C., giving a gold scarab signet to Huy (the owner of the tomb) and investing him as viceroy of Ethiopia with this gift.

The beautiful hard-stone scarabs which the ancient Egyptians wore were engraved by a lapidary, who cemented the gemstone to a bench with wax and resin and then engraved it by means of a bow drill. This jeweler's tool looks a little like a bow and arrow. The arrow part is the drill shaft. One end of it revolves in a hole drilled inside a hollow wooden cap, which the lapidary holds in his left hand. The string of the bow is twisted around this shaft, and the engraver

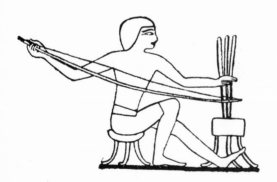

Egyptian jewelers engraved ornamental scarabs by means of a bow drill, shown here. The string of the bow, moved back and forth, oscillates the drill shaft.

moves the bow back and forth to make the drill revolve. The "arrow head" of the shaft is in fact a tube of soft copper. This copper tip is smeared with natural emery, a grit hard enough to cut through the hard gemstone.

The Greeks and Romans acquired the technique of gem cutting from ancient Egypt, and used it to make their own gem signets, or seals. Because most Greeks and Romans thought insects attractive to listen to and watch, some of these signets portray insects. Occasionally the kind of insect depicted on a gem lets us guess what kind of person the owner was. Thus a carnelian engraved with a cicada could have belonged to an ancient Greek musician, someone who played the lyre. The cicada was the symbol of Apollo, the god of music. A butterfly symbolized the soul. A lover might present his sweetheart with a signet engraved with a butterfly to show that he was hers, body and soul. On the other hand, a gem engraved with a grain of wheat supported by two ants might have been worn on the finger of someone who was a keen farmer. The ant was sacred to Ceres, goddess of agriculture, because it had taught man how to store grain in underground granaries—something ants themselves were very good

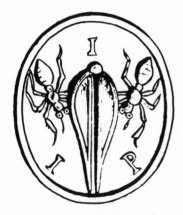

This Roman seal, a grain of wheat between two ants, probably commemorates the belief that ants taught men to store grain in underground granaries.

Locust, symbol of amatory powers, appears on this Greek seal—perhaps worn by someone courting a woman.

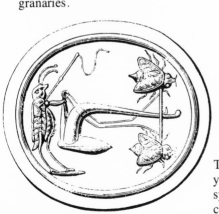

This grasshopper plowman, driving a yoke of cicadas, may have symbolized the charm of farming and country life.

at doing. A gem depicting a locust would probably have been worn by a great lover of women. We know from Greek inscriptions that the locust was symbolic of amatory prowess. Locusts were also used as an aphrodisiac. Another signet, showing a grasshopper driving a yoke of cicadas that pull a plow might have been the kind of gem commissioned by an owner who was both a poet and a farmer, as were Virgil and Horace. Very much later the Emperor Napoleon I adopted the bee as his heraldic device, because of its tireless industry.

Another kind of insect gem that begins to appear during Roman times was the fossil insect in amber. Amber is the fossilized resin formed on trees of the forest that covered most of northern Europe 40 million years ago. These trees included, among others, oaks, pines, poplars, walnuts, aloes, cinnamon, camellias, tulip trees, camphor trees, and redwoods like the Californian variety. Trees that were affected by disease or accident, or were punctured by tree-boring insects, oozed out an abnormal flow of resin over their torn bark. Insects wandering over such a sticky patch were caught and engulfed in succeeding flows of resin. The wounded tree soon died and fell to the ground, and the fallen tree trunks became embedded in the earth of the forest. A geographical upheaval transformed the forest into a seabed, and in time the resin-coated insects became fossils. Storms detached pieces of amber from the bottom of the sea. In the Middle Ages it washed ashore in such quantities on the Orkney Islands, north of Scotland, that the inhabitants lighted fires of it.

The ancestors of the present-day Anglo-Saxons and Scandinavians, however, found that there was a good demand for amber from Roman traders. The Roman aristocrats liked to carry a polished ball of amber around with them during the hot days of summer. It made their hands feel cool. The ancient Greeks had made beautiful carvings out of amber found in Sicily and on the shores of the Adriatic. The Romans admired them, as they admired everything Greek. They were fascinated by the insects which they could see through the polished, transparent amber. Every educated Roman would have echoed the words of the English poet of the eighteenth century, Alexander Pope:

> *Pretty in amber to observe the forms*
> *Of hairs, or straws, or dirt, or grubs, or worms.*

These things, we know, are neither rich nor rare,
But wonder how the devil they got there.

The Roman poet Martial also wrote a poem about amber that had an ant embedded in it:

While an ant was wandering in the shade of the poplar tree,
A drop of amber enfolded the little creature.
So that although in life it was regarded as nearly worthless,
By its manner of burial it has now become precious.

After the end of the Roman Empire, artists forgot about insects, with only a few exceptions. Odd pictures of insects occur in medieval illuminated manuscripts, such as an eleventh-century Exultet Roll from Gaetna in Italy, which shows bees gathering honey. This neglect of insects went on till the seventeeth century. When Thomas Moffat began to gather material for the first book about insects published in English, he and his collaborator Thomas Penny spent more money on the illustrations than on the text. Then his collaborator died, and an executor tore up all the illustrations by mistake. Moffat had to get them repaired at enormous expense.

It was left to a woman to bring life back to insect painting. Mademoiselle Sybille de Mérian had been born in Basel, Switzerland, the daughter, sister, and mother of celebrated engravers. Though she was married, she is always referred to as Mademoiselle, because in her day—in the seventeenth century—only ladies of title could be called Madame. Her marriage did not bring her happiness. Her husband lost all his money, then abandoned her. She joined a religious society and became interested in the idea of natural theology, or, as it was called at the time, "the theology of insects." Did the metamorphosis of insects form a sort of analogy to our life on earth and in the hereafter? After sketching flowers in Germany and in the tropical hothouses established in Holland, Mademoiselle de Mérian set out for Surinam, a Dutch colony in South America, so that she could paint insects in their native jungles. Fired by the books of Marcello Malpighi on the metamorphosis of the silkworm, she undertook a systematic study of the tropical insects of Guyana. As is well known, South America has the most abundant insect life in the world, and particularly has a wide variety of butterflies. At the age of fifty-four, Sybille began her two years' residence in the deadly fever-ridden jungles of Surinam. Alone, or accompanied by a few faithful Indian helpers, she wandered through the forests, painting insects in their various stages of development (caterpillar, nymph, and butterfly), together with the plants on which they lived and the animals and birds that ate them. In 1705 her great work, *Metamorphoses of Surinam's Insects*, appeared simultaneously in three languages. It has been said that it reestablished insect painting as a fine art in Europe.

Although neglected in Europe, insects had always been of compelling interest to the artists of the Far East, and they are still much more a staple of art with the Chinese and Japanese than they are with us. In China, the cicada had emerged as the most important insect amulet three thousand years ago under the Shang Dynasty. Amulets in the form of cicadas continued to be made down to the middle of the twentieth century, and a jade cicada was placed on the tongue of the dead in China. (The Chinese believed that jade would preserve the body, which of course it did not do.) The cicada was the symbol of rebirth into the next world; it was associated with speech, and therefore placed on the tongue. This custom is probably based on the fact that cicadas are never silent during the hot days of summer.

The use of cicada amulets for burials began as early as the Chou Dynasty (1122–255 B.C.), but long before, under the Shang Dynasty (1766–1122 B.C.), the insect had become an important art motif. Cicadas were carved on bone scepters and used to ornament bronze vessels. Later, under the Han Dynasty, (206 B.C. to A.D. 220) men began to carve jade buckles in the design of a mantis preying on a cicada, the symbol of destiny.

Besides being carved in jade and molded in bronze, insects must also have been portrayed in ink on silk or paper as early as the Han Dynasty, if only because there are so many allusions to them in literature. "If insects are remarked on by poets," says a manual for painters, "how can they be neglected in painting?" To the eye of the aesthetic Chinese, insects symbolized summer, particularly in association with the Seven Flowering Grasses of Summer, or with the flowers of spring and fall. Many painters made their name as portrayers of insects. In the T'ang period (A.D. 618–907), T'eng Wang-ying was renowned for his butterfly paintings. It was said that he was able to breathe life into them with the first stroke of his brush. Like most Chinese insect painters, he painted his flowers in color and his insects in black ink. T'eng Ch'ang'yu, Hsu Ch'ung-ssu, Ch'iu

Yu-liang, and Hsieh Pang hsien were all adept at painting bees and wasps as well as butterflies. Many other Chinese insect painters might be mentioned—including Chao Ch'ang and Kuo Shou-ch'ang, whose flying and hopping insects vibrated with life.

Almost everything imagined by the Chinese artist was to be developed by his Japanese counterpart. Japanese carvers of netsuke, small toggles used as fasteners, and *inro,* medicine boxes, labored to portray in ivory, wood, and lacquer the symbolic ideas of the Chinese about insects and to add others of their own. Thus the dragonfly (*tombo*) symbolized Japan, because the emperor Jimmu Tenno had once climbed a lofty mountain peak and remarked that the island chain of Japan "was like a dragonfly licking its tail." Another

The dragonfly, depicted gracefully in this old Japanese print, symbolized Japan, because the island chain looked like a dragonfly licking its tail.

A boy chases dragonflies. If he belongs to the warrior class, he can wear a dragonfly amulet when he grows up.

emperor, Yuriaku, was bitten by a gadfly, which was then eaten by a dragonfly. Though the dragonfly looks like such a sedate and placid insect, it is really a ferocious and predatory hunter of other insects, and so became the symbol of martial glory in Japan. Those boys whose birth had equipped them with the right social background to take part in warfare wore dragonfly amulets.

The ant and the bees were symbolic of industry and thrift. The butterfly symbolized the soul, summer, and marriage. At marriage ceremonies two paper butterflies were placed together as part of the marriage service, with the male on top, just to show who was boss. The butterfly was also used as a comparison by Japanese men for the young, frail, light-hearted, frivolous, and pleasure-loving *mousmé*, or teenage girl of Old Japan, always flitting in fickle fashion from one beau to another.

All these symbolic ideas were worked out to the full by the Japanese craftsman, who would, for example, decorate a staghorn pipestem with tiny bronze ants moving in procession, or portray glowworms in a cloud of gold lacquer. The Japanese armorer even constructed helmets that looked like beetles. Some artists working in metal devised the cruel but ingenious process of encasing a live insect in clay. The clay was calcined, the fragments of the insect removed, and then an exact cast was made of it in bronze or some other metal.

We have looked at the use of insects in the arts and in the inspiration they provided for artists. But insects have been no less important in providing colors and other artistic raw material to the craftsman. The finest scarlet has always been produced by insects. This color, which is used even today to dye the coats of English foxhunters and the uniforms worn by the Guards regiments in the British army, is made from the cochineal insect, or *Dactylopius coccus*. The coccus was originally a native of Mexico, like the firebug, and it was valued so highly by the Aztecs for dyeing their gorgeous cotton dresses and robes that taxes were paid to Montezuma in cochineal insects.

This dark, brown-red insect, about the size of a pea, lives on the prickly-pear cactus, or nopal. The original dye was a dull crimson. Then a chemist called Kuster discovered that this dull red could be turned into brilliant and exciting scarlet by treating it chemically. The first shipment of cochineal reached Europe in 1523. At first everyone thought cochineal was a plant seed. Then someone demonstrated that it was an insect by throwing it into hot water so that it opened out. From that moment many people in Europe began to look for an alternative source for the cochineal insects, or to plan to smuggle the insect itself out of Mexico. The cochineal became the third most important commercial insect of all time, next to the silkworm and the bee.

The Mexican Indians reared the coccus by setting up a prickly-pear cactus plantation, or nopalry, which was protected from the west

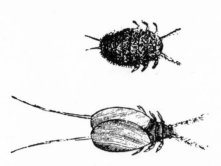

The cochineal insect, a creature resembling a mealybug, was so important a source of red dye that Aztecs paid their tribute to Montezuma in these creatures.

Cochineals live on the prickly-pear cactus, from which they were removed with a squirrel's-tail brush. It took 70,000 of them to make a pound of dye.

wind and hedged in with reeds. Prickly-pear cacti (*Cactus tuna*) would be planted from cuttings, and a stack of newly hatched cocci acquired; the insects would be kept for a time indoors and later moved outdoors in an open-sided shed. In August and September the pregnant females would be gathered and placed on the plantation of cacti to breed. In four months the first gathering would begin, followed by two more. The Indian gatherers would brush the insects from the cacti with a squirrel's tail. The Indian women (men were not employed on this tedious task) would have to squat for hours beside one plant before it was cleared. The cochineal insects were then killed with boiling water, dried, and sent to Europe in boxes for processing. It took 70,000 insects to make a pound of cochineal.

Other countries were eager to get hold of these valuable insects. A Frenchman named Thierry de Menouville smuggled out some, still on their plant food. He only got them as far as Haiti, where a revolution broke out and all his cocci were lost. However, he discovered a Haitian variant of the insect, which lived on a different cactus. In 1806 another French insect smuggler, M. Souceylier, a French naval surgeon, tried again. This time the cocci got as far as Toulon, where they were handed over to the professor of botany in the Botanic Gardens. The cochineal insects did not find the French prickly-pear cacti as nourishing as those of Mexico. They died of starvation. Finally the cochineal insect was introduced to the Canary Islands by the Spanish, who by now had lost Mexico and had no further interest in maintaining the Mexican monopoly of cochineal.

The Canary Islanders are painstaking farmers, as they have to be to survive on their volcanic-soil islands, where rainfall is scanty. Conservative peasants, they at first regarded the cochineal insect with profound distrust. What happened next can be told in the words of the Victorian savant, Dr. T. L. Phipson:

> Who would have thought in 1835 that the years of the grape-vine of Tenerife were numbered? Then Tenerife had been a vine producing country for three hundred years, and when a gentleman introduced the cactus and cochineal there from Honduras, he was looked upon as an eccentric man, and his plantations were frequently destroyed at night. However when the grape disease broke out (*Phylloxera vasatrix*) Tenerife was gradually forsaken by vessels in quest of wine which could no longer be supplied, and with starvation staring them in the face, the inhabitants turned to cochineal growing; wherever a cactus was seen upon the island a little bag of cochineal insects was immediately pinned to it. The plan succeeded admirably. An acre of the driest land planted with cactus was found to yield three hundred pounds of cochineal, and under favorable circumstances, five hundred pounds, worth £75 to the grower. Such a profitable investment of land was never before made.

By 1831 the French had managed to establish the coccus in Algeria, North Africa, where it became extremely important for the economy. It was cultivated in fourteen nopalries. England in 1855 imported 1,400 tons of cochineal, valued at £700,000. Shortly afterward the bottom dropped out of the cochineal market. New aniline dyes had been discovered. But even today, as has been seen, cochineal continues to be used for fine dyeing and also for making carmine and crimson-lake watercolors. This useful insect had done mankind an inestimable service in providing an edible dye.

Before we look at two more members of the coccus family and the colors they provide, let us just glance at the most important pigment insect of the Pacific, the *awets* caterpillar. We have already mentioned that the Maoris of New Zealand painted themselves for certain occasions. Their permanent body decoration was provided by tattooing. In early days, every Maori male of high rank was heavily

A French colonist in Algeria explains to an Arab visitor how cochineal is concocted from the cacti.

tattooed. Women were only allowed a few small tattoos, partly because the whole business was extremely painful. The pigment, smeared on tiny adzes made from bird bone, was chiseled into the face, with blood spurting out at every stroke. The colors used were powdered wood or resin, and a caterpillar called *awets*. Only those

The Maoris of New Zealand tatooed themselves heavily as a permanent body decoration, employing a dye made from a fungus-afflicted caterpillar called the *awets*.

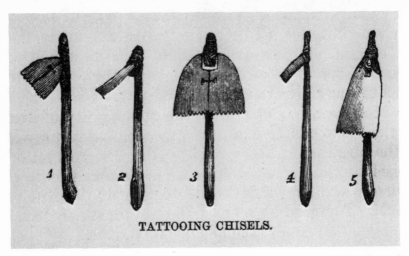

TATTOOING CHISELS.

Using tiny chisels and adzes, made from bird bones, the Maoris implanted pigment in the flesh of the face. An enemy's finely tattooed head was especially prized among spoils of war.

caterpillars were used that had been invaded by a fungus, which killed it and took over its body, turning it into a rich, cherry-red color. The Maori tattooist improved this color by charring the caterpillar. The resulting product was a fine dark dye, highly valued because it was difficult to obtain. The Maoris not only tattooed their faces, they also preserved the heads of slain enemies by smoking them. Many a man lost his head because his opponent liked the fine facial tattoos he wore.

Coccus ilicis is a close relative of the cochineal insect. It used to provide an important insect dye, which has dropped out of use. This coccus lives on the kermes oak, a shrub common on the dry arid wastes of Mediterranean France and Algeria. The dried insects, wrapped in their cocoons, were sold under the name of "grains of kermes." The dye made from kermes was used to dye carpets and morocco leather, and it gave its distinctive color to the red fezzes worn by men in eastern Mediterranean countries, such as the Zouaves, Algerian troops in French service. They wore a uniform of brilliant colors, which was so dashing that it was adopted by a number of militia units and appeared in some early battles of the Civil War.

The *Coccus laccae,* or *Laccifer laccae*, is the insect that produces the lacquer of South East Asia (not to be confused with that of China and Japan, which comes from a tree). This insect lives on many Indian trees, such as the Indian fig tree, the pagoda tree, the jujube, and the *Croton lactiferum,* as well as on the *Butea frondosa* and others. These trees, especially the Indian fig tree, are noted for their milky sap. The *Coccus laccae* pierce the bark and the sap oozes out, producing first a protective layer, through which the insect tunnels, and subsequently a resinous tomb for the mother insect and a depository for her eggs. This resinous deposit is broken off the tree and processed into shellac. Shellac is a natural dyestuff that can be used as both a varnish and a stiffener. In the nineteenth century, it was used to make sealing wax, a use that has continued down to our own day. It was also used to stiffen women's silk hats and to varnish everything from pictures to railroad cars. In the twentieth century, before plastic came into use, phonograph recordings were prepared from a shellac formula.

Magnificent lacquer work is made from the shellac produced by Southeast Asian trees, which is formed after an insect burrows into the bark to deposit its eggs.

Indian lacquer work from Karnul

Lacquered leg of a bedstead from Sindh

Indian
lacquered
boxes

Shellac was used to make lacquer paint. The lac was collected
from the trees in the jungle by the Bhils, a forest people of India, who
traded it to the lacquer craftsmen in the towns. The craftsmen mixed
it with colors or with tinfoil to make it look silvery. Finally it was
used to produce marbles, toys, bangles, and lacquered boxes or
furniture.

Several branches of art owe a lot to the gall wasps, *Cynips quercus
folii*. These insects, belonging to the family Cynipidae, bite the bark
off trees and inject it with an irritating solution, at the same time

laying their eggs in the opening. A ball of wood called a gall begins to form as the sap of the tree reacts to the irritant. In the middle of this gall the larva of the cynips, or gall wasp, begins to develop. When fully grown, the larvae bite their way out of the gall, leaving an exit hole.

The gall wasp's bite injects an irritating solution into the wood of a tree, which forms a swelling, or gall. From oak galls men have made the world's most permanent ink.

The solution injected by the cynips has important chemical properties. In early days it was put to use in tanning, especially in tanning morocco leather, which was treated with Aleppo galls. Another early use of galls was to make ink. Modern soluble inks for use in fountain pens are hardly more than paints, which rest on the surface of the paper and can be removed completely by an expert. (This helps the forger.) Ancient inks were a pigment and mordant combined, which bit right into the paper. Medieval ink was made from bruised oak galls, gum arabic, sugar, iron sulfate, and copper sulfate. So penetrating was this ink that even after all traces of it have been washed off the page, so that everything on the surface looks blank, parts of the ink that have sunk beneath the surface can still be read by means of an ultraviolet lamp. Francis Bacon once borrowed a monastic manuscript from its owner and then presented it to someone else, having first scraped off the owner's name with a knife. Three hundred and fifty years later, I was able to demonstrate that the manuscript actually had belonged to the person whose name was erased, and all because of the irritant solution from a gall wasp.

7

INSECTS AS FRIENDS AND TEACHERS

All insects," wrote French historian Jules Michelet in 1875, "teach certain noteworthy lessons." He went on to say that although insects could not really be domesticated, they did respond to kind treatment and recognized people who were good to them. History is full of insects that have helped people, unconsciously perhaps, by becoming their friends and comforters. They have cheered the loneliness of the captive, inspired the great to brave deeds, and helped the inventive discover many things that were to benefit mankind.

The best-known person to become inspired by an insect was Timur the Tartar. Timur, or Tamerlane, was the most forceful personality of the fourteenth century. He built up a world empire stretching from Russia to Arabia in one direction and from Turkey to India in the other. During one moment of his career, however, Timur had been defeated and forced to take refuge from his enemies in a ruined house. He sat alone for hours, plunged in despair. Then at length his attention was caught by a movement on the wall opposite to him. An ant was trying to carry something up the steep wall. Timur, who was a better strategist than an entomologist, says it was a grain of wheat, but it was probably a case containing eggs. At any rate the burden was bigger than the ant itself. To distract his mind from his own troubles,

Timur began to observe the creature's movements. The ant tried to climb the wall sixty-nine times. Each time it fell to the ground with its burden. Undaunted, it persevered, and on the seventieth occasion, it succeeded. "That sight," says Timur in his *Memoirs*, "gave me courage at the moment, and I have never forgotten the lesson it conveyed." He left the ruined hovel to become the conqueror of Asia.

A much more extraordinary story than this is told by Frank Cowan of a certain Frenchman named Quatremer Disjonval. Though French by birth, Disjonval lived in Holland, and just before the French Revolution he took part in the revolt of the Dutch republicans against their king. This rebellion was suppressed with the help of a Prussian army, and Disjonval was condemned to twenty-five years of imprisonment in a Utrecht dungeon. He remained there for eight years, during which time his sole companions in solitary confinement were the spiders that shared his cell. As he watched them, he began to discover that they were very sensitive to approaching changes in the weather. By careful observation of their behavior, he was able to foretell what the weather would be like ten to fourteen days in advance.

By now the French Revolution had broken out, and in 1793 the French revolutionaries announced that they were prepared to help uprisings elsewhere in Europe. In 1794 a French revolutionary army invaded Holland. Holland was then much more enclosed with its water ways and branches of the Zuyder Zee than it is now, and the French army—horse, foot, and artillery—kept pushing on over the frozen sea, canals, and rivers. Then, all of a sudden, cracks began to appear in the ice. A sudden and unexpected thaw, very unusual for that time of the year, early December, had set in. The French decided that they must hastily withdraw or be cut off. Their commander opened negotiations with the Dutch government.

Disjonval had learned all this through chatting with his jailer. He realized that if the French fell back, he might have to wait another eight years to get out of jail. So he bribed his jailer to send a letter to the French commander. In the letter Disjonval told the French general that there would be a severe frost within fourteen days, information he had gathered from his spiders. The French leader believed Disjonval's weather forecast and pressed on. The cold weather arrived in twelve days, bringing such a severe frost that the rivers and canals froze hard enough to bear the heaviest artillery. On January 28, 1795, the French army entered Utrecht in triumph, and Quatremer Disjonval was released from prison.

About a century before Disjonval, another French prisoner also learned something interesting about the spiders in his cell. Frank Cowan describes his experience thus:

> Monsieur de———, a captain of the Regiment of Navarre [in the army of Louis XIV] was confined six months in prison for having spoken too freely of M. de Louvois [one of Louis' ministers] when he begged leave of the government to send for his lute to soften his confinement. He was greatly astonished after four days to see, at the time of his playing, the mice come out of their holes and the spiders descend from their webs, who came and formed in a circle round him to hear him with attention. This at first so much surprised him that he stood still without motion, when, having ceased to play, all those spiders retired quietly into their lodgings.

A century later, in the 1820's, another lonely prisoner was consoled by insects, this time in Italy. Silvio Pellico had been incarcerated in a very unpleasant prison for having tried to promote an Italian nationalist revolution against the Austrians. (He was probably innocent.) Pellico wrote a book about his life in jail, *My Prisons,* which became one of the best-sellers of the nineteenth century. In America and England, young girls, using it as a textbook for learning Italian, cried over it, and Pellico's work became the best-possible piece of propaganda for Italian nationalism. With both England and France backing Italy's nationalist movement, Italy finally became a united country in 1870. Of his solitary confinement in Venice, Pellico wrote:

> I so rarely saw human beings that I began to pay attention to some ants who came onto my windowsill. I fed them well. They went off and returned with an army of their companions. Soon, my casement was covered with these little creatures. I also became interested in a lovely spider which spun its web on one of the walls of my cell. I fed it with gnats and midges, and it became so friendly that it came onto my bed and into my hand to seize its prey from my fingers.

Mention of spiders cannot fail to introduce (especially in a book written by a Scotsman) some mention of Robert the Bruce. Scotland, an independent kingdom in the north of the British Isles, had imprudently called in Edward I of England to sort out a tangled dynastic succession to the Scottish throne. Edward had used the opening thus given him to turn Scotland into an English satellite. The Scots revolted but were crushed again and again. Not merely were they overwhelmed by superior numbers of English, but the army of Edward carried a new and deadly weapon—the longbow.

Then Scotland found a new leader in the person of Robert Bruce. Bruce had himself crowned at Scone as King of Scots. Edward, stung by this act of defiance, attacked Robert, driving him first to the wilds of Athole, then out of Scotland altogether, to the little isle of Rathlin, off the north coast of Ireland. There Robert lay hidden, and everybody supposed him to be dead. One day, lying in bed, Robert noticed a spider, which had begun to spin a web over his head and was trying to reach a beam on the ceiling to attach the next thread. Six times the spider tried, and six times it failed. "Now," said Bruce to himself, "shall this spider teach me what I am to do, for I also have failed six times." The spider made a seventh attempt—and succeeded.

The following spring, Bruce left the island, collected three hundred followers in Scotland, landed at Carrick, and at midnight surprised the English garrison at Turnberry Castle. He went on to win victory after victory over the English until in 1314 he delivered to them a decisive defeat at a place called Bannockburn. Thereafter the Scots remained independent of the English until 1603, when their positions were reversed—a king of Scotland, James VI, became king of England.

A survivor of the Wyoming Massacre, during the American Revolution, crawls into a hollow tree to escape Indian pursuers. A friendly spider, by spinning a web over the hollow, helped conceal the hidden man.

A spider also once saved Frederick the Great of Prussia. One day, at his palace of Sans Souci near Berlin, Frederick walked into his anteroom to pour himself a cup of chocolate, which he usually drank at that hour. He filled his cup from the tiny pot, then remembered that he had left his handkerchief in his bedroom and went back to fetch it. On his return he found that a great spider had fallen from the ceiling into his cup. He rang the bell for a fresh pot of chocolate. The next moment a pistol shot sounded in the pantry. His cook, who had been bribed by the king's enemies to poison his chocolate, thought his treachery had been discovered and shot himself. Frederick had a spider painted on the ceiling of the room as a reminder of his deliverance.

More than one fugitive has escaped from a pursuing enemy because he crept into a hiding place and an industrious spider instantly set to work to cover the entrance to his place of concealment with a web. The pursuer arrived on the scene, saw the web, was convinced that nobody could have hidden there, and passed on. This adventure happened to King David of Israel, to Muhammad, while being pursued by the Koreishites during his flight from Mecca, to Saint Felix of Nola, and to one of the survivors of the Wyoming Massacre, who crept into a hollow tree and was concealed from his Indian pursuer by a spider's web.

The great men of history have always had an eye for little things, including insects. This can be seen from what is perhaps the strangest of all insect stories, one which concerns a Danish prince of the Middle Ages called Hamlet—the real-life inspiration for Shakespeare's hero. Hamlet—or Amleth, as his name was spelled in Danish—was the son of the king of Denmark, but his father had been murdered by his wicked Uncle Fenge (a good name for a villain), who then married Amleth's mother, Gerutha, or Gertrude. At this point Fenge would now have made a clean sweep and murdered Amleth as well, but for the fact that the young prince cleverly feigned madness. He crawled about the ashes of the family hearth, talked wildly about getting his revenge for his dead father, and spent his time making—of all things—tent pegs. If Amleth really was mad, Fenge thought, there was no point in putting him out of the way. But what if the young prince was just shamming? Fenge arranged a test. Amleth would find himself in the forest, apparently alone, with a beautiful young girl. But Fenge's spies, hidden in the bushes, would see everything and report on what the prince did. Surely, left face to face with a lovely maiden, Amleth would do or say something to betray himself. Unknown to the usurper Fenge, however, one of his spies was secretely devoted to Amleth. This man, Amleth's foster brother, wondered how to send Amleth a message that trickery was intended. Since it was impossible to shout to Amleth without betraying himself, the foster brother caught a gadfly, tied a straw to its abdomen, and released it in Amleth's direction. The gadfly flew past where Amleth was sitting with the young girl. The prince noticed its strange attachment, and instantly realizing that this was no accidental encounter, did not give himself away to the spies.

Eventually, when the right moment came, Amleth got his revenge. Having been sent into exile in England, Amleth spread the rumor that he had died. Then he returned to Denmark in time for his own funeral feast, which the king and court were celebrating with great enthusiasm. Amleth waited till the last of the revelers had dropped from the mead benches dead drunk, then tore down the tapestries from the walls and threw them over the sleepers. Now at last the purpose of the tent pegs became evident. Amleth removed them from their hiding place in the hall and nailed down the tapestries on top of the drunkards. When he had run through his uncle in his bedroom, all that remained to do was set fire to the hall. "Thus Amleth," wrote the Danish historian Saxo Grammaticus, "earned everlasting praise, for he hid wisdom under a self-made cloak of folly, and with cunning and valor not merely revenged his father, and won back his kingdom, but saved his own life." All Amleth's cunning and valor, however, would have availed him little but for the fact that flying insects can carry more than their own weight.

The same habit of observing insects that had saved Amleth preserved the Abbé Latreille, a French entomologist who tells us the story himself in his *Histoire des Insectes*. Latreille was able to classify a particular genus of beetle, which he named *Necrobia*. This strange name, which is Greek, means "life in death." It was given to the genus by Latreille to commemorate how one of them saved him from certain death.

In 1792 Latreille had only just begun to study insects. He had written a few papers on the subject, but he was kept busy by his job, that of being a country priest at Brives-la-Gaillard, in the province of Limousin, in France. At this time the French Revolution was under way. The revolutionaries had begun by reorganizing the French state, now they planned to reorganize the Catholic Church in France in such a way as virtually to destroy its identity. Priests were, in effect, to become civil servants of the government and were to take an oath of allegiance to the state. Many French ecclesiastics felt that they could not, in good conscience, take this oath, even though defiance of the government order might mean execution by the guillotine. Latreille was one of them. He and several others were arrested. But because many Catholic Frenchmen would have objected vociferously if these priests had been sent to the guillotine, the Revolutionary Tribunal sentenced them instead to what was called "the dry guillotine": imprisonment in the convict colony of Guyana, in South America, part of which was the infamous Devil's Island. We can imagine the unfortunate priests, jolting across France in farm carts, under the guard of soldiers from the *armée revolutionaire*, wishing they could get a shave, or a change of clean linen, till they reached Bordeaux, the great seaport from which they were to sail to South America.

Once in the port, they were locked up in the prison of the Grand Seminary till a ship should be ready to take them to Guyana. While they waited, a doctor called to attend to a sick prisoner, a bishop. As the doctor was chatting with his patient, he noticed Latreille suddenly make a pounce on a beetle that had just emerged from a crack between the floorboards of the prison, examine it, and pin it to a mounting board. "Is the beetle a rare one?" he asked Latreille. "Yes," was the reply. "Then perhaps you might like to give me the beetle as a present for a friend of mine," said the doctor. "He is a young man called Bory de Saint Vincent. Though young, he has a large collection of insects, and he is very interested in classifying them. I expect he will be able to classify your beetle." Latreille agreed. He gave the doctor the beetle and asked him to get Saint Vincent to classify it, and tell him the species.

At his next visit the doctor reappeared with the beetle and a crestfallen expression. Monsieur de Saint Vincent, he told the abbé, had been unable to classify the beetle (which, as you will have guessed, was a *Necrobia*, the *Necrobia rufficollis*), and so he could not tell the prisoner what species it was. In fact, Bory de Saint Vincent was a real naturalist, one who knew all about insects, one who would have read the abbé's scientific papers about them. The jailers at the Grand Seminary did not allow their prisoners pens or paper, but Latreille gave the doctor a verbal message for Saint Vincent. "Tell him," he dictated to the doctor, "that I am the Abbé Latreille, and that I am going to die in Guyana, without having published my book, *Classification of Insects*."

Alerted by this message, Bory de Saint Vincent determined to save this fellow entomologist. He used his influence with the authorities to get Latreille out of jail on the pretext that he was too ill to make the voyage. He stood bail with his uncle for the abbé's good conduct, and the prison ship sailed for Guyana without the Abbé Latreille on board. It never reached its destination, because it foundered off the coast of Spain and all the prisoners went to the bottom. Meanwhile, back in France, Bory de Saint Vincent got Latreille's name struck out of the list of exiles, and the *Necrobia rufficollis* got its name.

One very positive way in which insects have befriended men is by giving them ideas for new inventions. Clever men have never been above getting a good idea from our tiny friends, even though they have not always been honest enough to acknowledge the source of the idea. Once you begin to work through the list of major discoveries, it is remarkable how many of them were in fact suggested by insects.

One of the most ingenious inventions of the nineteenth century was the tunneling shield. This was a circular iron plate which was forced into soft ground at the head of a tunnel, while bricklayers bricked a circular wall inside it. The invention of this tunneling shield helped the two famous British engineers, Sir Marc Isambard Brunel (1769–1849) and his son Isambard Kingdom Brunel (1806–1859), to construct the first tunnel under a navigable river, the Thames. The terrain through which the tunnel had to be dug, sand and gravel, was so treacherous that the project, begun in 1825, was only finished in 1843. At one point the roof caved in and the workmen, including Isambard Kingdom Brunel, had to run for their lives. Young Brunel, after plugging the roof of the tunnel with bags of clay dropped into the hole from a barge, recalled the strange cutting mechanism of a creature he had studied.

Nowadays this creature would be classified as a shellfish, but in those days it was called the shipworm. The shipworm, or *Teredo navalis*, was a menace to shipping, boring holes into the wooden bottoms of ships by revolving the shell plates in its mouth. Brunel simply copied this mechanism, substituting metal plates for the shell ones. By pushing forward the shield, the tunnel could be bricked over almost as fast as it was dug, thus sealing off possible leaks. Eventually the tunnel was finished, and it is still in use for underground trains today.

Brunel was modest enough to acknowledge his debt to a worm, but not so the Chinese inventor T'sai Lun, who benefited from a wasp. According to legend, this inventor, a prominent government official under the emperor Ho Ti (A.D. 89-106) first made paper out of tree bark, hemp, or fish nets, and later out of vegetable fibers. Before the invention of paper, the Chinese had written on bamboo slips, which were troublesome to prepare, or silk, which was expensive. The Chinese papermakers collected bark, grasses, or other plants, pounded them with a mortar till the fibers separated, boiled the fibers in lime, and washed the resulting pulp clean. The fibery pulp was then dipped out of a vat on a bamboo sieve, tipped out onto cloth, and finally dried in the sun.

Not till the eighteenth century, and after twenty years of research, did a French entomologist demonstrate that what T'sai Lun was supposed to have invented, wasps had already been making for countless years. René Antoine Ferchault de Réaumur (1683–1757),

Réaumur, the eighteenth-century French scientist, was the first European to discover that wasps' nests were made of paper. Watching these insects at work probably inspired the Chinese invention of the paper-making process. The latter simply substituted a lime solution for the insects' natural secretions.

a French naturalist and physicist, one day noticed a wasp on his windowsill, busily gnawing away at the wood. He saw the insect detach a bundle of fibers and collect them with its feet; later he discovered that these fibers were as thin as a hair and about one tenth of an inch long. He was subsequently able to observe the wasps as they moistened the fibers with an internal secretion, then rolled them into a ball, and spread them out to dry.

Another Frenchman, Jules Michelet, described the wasps at work:

The wasps suspend their edifice in the air, and built it out of a strong,
coarse paper, to obey the heaviest rains. To make this paper, they
hasten to the forest, where they select some thoroughly prepared
wood, which has been long soaking, and has been already steeped by
nature, just as we steep flax. Then they gnaw, tear, loosen and sever
the rebellious filaments, pound them into pulp as we do the linen rags,
and knead them with a heavy tongue. After the paste has been mixed
with a viscous and adhesive saliva, it is spread out into thin layers.
With teeth closed like a press, the work is completed.

Obviously, the only major difference between the wasp papermaker and the Chinese papermaker is that the latter had to
substitute the solvent action of lime for the natural secretions of the
wasp. It is most likely that T'sai Lun, or whoever did invent paper in
China, simply watched wasps at work.

The Chinese were also extremely insect conscious. Chinese
children were taught not to step on industrious insects, such as bees,
ants, and so on. As early as the third century A.D., the nests of a
hunting ant (*Oecophylla smaragdina*), were sold to farmers near
Canton so that they could place them in their citrus groves and let the
ants destroy citrus insect pests. The habit of close observation of
insects, which the Chinese practiced, may help to explain why

Nature's insect balloonist, the spider,
spins four filaments and prepares to
fly with them. In flight, it controls
direction by lengthening and
shortening the threads.

another insect invention was taken up by them, kite flying. In 1885, Henry C. McCook, an American entomologist, gave a classic exposition of insect kite flying in a book called *Tenants of an Old Farm*. He describes a small saltigrade (leaping) spider sitting on the rail of a fence, spinning filaments of silk. The filaments hardened as soon as they were spun and then floated up into the air. Eventually four threads were spun, and the spider then let out enough of it that the buoyancy of the silk would overcome the insect's weight. When this happened the spider simply took off into the air. Once airborne, it controlled its flight, as a kite flier does his kite from the ground, by lengthening or shortening the threads. Kite flying became one of the great pastimes of the Far East. In Japan military observers were even sent up on monster kites to spy out the lie of the land.

A master engineer, the spider thinks nothing of spinning a suspension bridge to carry him over a stream.

Another invention which is thought to have been suggested by the spider's spinning action is the suspension bridge. McCook described how an orb-weaving spider throws a line across a brook, fixes its anchorage to clumps of grass on one bank, fastens the other end of the line to the abutment on the far bank, stretches a foundation line, then another below it, and finally constructs trusslike supports between the two.

Yet another invention was suggested to mankind by the spider, the diving bell. Before 1819, when Augustus Siebe invented the diving suit, the only way to work below the surface of the water, except by free diving, was to imitate the water spider. This creature, *Argyroneta aquatica*, weaves for itself a dome of silk, which is open at the bottom but closed at the top and watertight. In this natural diving bell, the spider descends, trapping a bubble of air within the dome. It can stay under water as long as its bubble of air lasts. The diving bell

The water spider spins itself a dome that is open at the bottom but closed at the top, thus creating a diving bell long before man thought of it.

was simply the spider's dome enlarged many times. It was bell-shaped, made of wood bound with iron, and inside there were seats for divers. The bell was lowered slowly through the water to the bottom. Water did not fill the bell, because it could not displace the air it contained, which became more compressed the deeper the bell sank. Obviously the divers could not go on breathing this trapped air for too long. Nor could they do much inside the bell. What they did do was to sally out of the bell, holding their breath, and attach ropes to objects lying on the bottom such as cannon, which could be salvaged by the mother ship. It was presumably with a machine such as this that Sir William Phips (1651–1695) was able to recover treasure worth at least a million dollars at present-day value from a wrecked Spanish ship in the Bahamas. It was largely owing to this successful start in life that Phips was knighted, made colonial governor of Massachusetts, and arrived in the colony in time to suppress the witch trials at Salem.

The ancient Greeks were disposed to credit the spider with the invention of weaving. According to an ancient myth, a woman named Arachne ("spider") changed into an insect because she dared to compete in weaving with the goddess Athene. Certainly there is something very suggestive of weaving in the way a spider lays down a framework of threads for her web, then connects them with others.

Another insect invention, said the Greeks, was that of storing grain in underground granaries. This is what the ant does, and many peoples of the ancient world, and even of modern North Africa, have followed the ant's example. It was accidental discovery—and confiscation—of an underground cache of Indian corn that kept the Pilgrims alive during their first winter at Plymouth.

Other ant inventions include plastering and the preparation of fine earth for walls and ceilings. Several species of ants show the greatest care in selecting gradings for their dwellings. Termites, for example, use such fine earth that it has been borrowed to make jeweler's molds. In the nineteenth century, Scots Presbyterian missionaries in Blantyre, Malawi, were at a loss to find suitable clay from which to make bricks to build a church. At length they tried the earth of termites' nests. All the bricks used in the imposing church at Blantyre—a building which is as large as some cathedrals—were made from earth quarried from the termite mounds that are such a distinctive feature of Central African scenery.

The extent of the indebtedness of human inventors to the insects is boundless. Take the honeycomb shape or hexagon. When in the sixteenth century the Japanese began to hear stories about European castles from the missionaries who had arrived in Japan from Portugal and Spain, they wanted to imitate them. But the would-be-castle-builders found themselves faced with an apparently insoluble problem. Earthquakes are extremely common and severe in Japan. How could the Japanese erect castles that would not immediately be shaken to the ground by the first shock? Once again the insects supplied the answer. The builders, who had obviously experimented with that clever insect device, the hexagon, used both by bees and wasps, made their foundation stones hexagonal. On a sort of mound of hexagonal-shaped rocks they erected a comparatively light and flimsy wooden castle. When an earthquake tremor arrived, the hexagon boulders rolled slightly in unison, like the wheels of a machine turning together. The wooden superstructure of the castle swayed but did not collapse. Did the castle architects get the technical information about the hexagon from the Christian missionaries? Well, the first important castle, that of Oda Nobunaga (at Azuchi) was apparently surmounted by a crucifix. The keep of a Japanese castle is still called by a name that suggests Christian influence. Be that as it may, the missionaries certainly gave the Japanese technical advise on another aspect of insects—the mosquito net. It has been used in Japan continuously ever since.

Right through history, insects have provided a perpetual source of inspiration for the inventive. In 1802 William Paley (1743–1805), an English scientist, predicted that one day an invention would be made, based on the mode of traveling used by a particular insect:

> Some years ago, a plan was suggested of producing propulsion by reaction. A stream of water was to be shot out of the stern of a boat, the impulse of which stream upon the water in the river was to push the boat forward. Now if naturalists are to be believed, this was exactly the device which Nature has made use of for the motion of some species of aquatic insects.

There were no airplanes in Paley's day, so he could not envisage that the kind of "propulsion by reaction" that he was dreaming of would one day be applied, not to water, but to air—in the jet plane invented by Sir Frank Whittle.

INDEX